Geological Landscapes of Britain

Julian Ashbourn

Geological Landscapes of Britain

Springer

Julian Ashbourn
Berkhamsted
Hertfordshire
United Kingdom

ISBN 978-90-481-8860-4 e-ISBN 978-90-481-8861-1
DOI 10.1007/978-90-481-8861-1
Springer Dordrecht Heidelberg London New York

Cover illustration: Figures from this book, pp. 20, 24, 25, 39 and 54.

Printed on acid-free paper

Springer is part of Springer Science+Business Media (www.springer.com)

Prologue

Geology is a fascinating subject, but those who have yet to embrace this fascination might be forgiven for thinking it is just a study of rocks. Inanimate objects which happen to be lying around under our feet. Natural historians and biologists may study all manner of flora and fauna, down to cellular level, together with their evolution and respective lifecycles. Archaeologists may revel in past intrigues and forgotten civilisations, always looking forward to a new and exciting discovery which might forever change our understanding of the past. Palaeontologists may immerse themselves in the fossil record, painstakingly assembling a complex jigsaw puzzle of life which, as it takes shape, opens their eyes to ever more wonderful discoveries. Oceanographers have a wonderful world of marine life and interesting physics to delight them. Environmentalists may occupy themselves with an appreciation of habitats and complex relationships. But geologists? Geologists study rocks.

Of course, the reality is that geology is every bit as fascinating and absorbing as any of the natural sciences, incorporating elements of most of them. There is the historic context, the remarkable area of plate tectonics and the birth of continents, the relationship with both the atmosphere and biosphere, the wealth of minerals and composition, and the unequivocal fact that we are dealing with the very fabric of our world. A fabric which is volatile, which is subject to amazing transformations and which underpins our entire existence. Far from being simply a study of rocks, geology represents a study of our world. Its creation, its composition, its past and its future. Furthermore, it is a study which reveals nature's most astounding creations, in all their beauty and

Fig. 1 Looking across the loch at Foyers in the Scottish Highlands

majesty. Geology is all around us. We can reach out and touch it almost anywhere. We can admire it from a distance, or examine it under a microscope. Once awakened, our fascination with this most interesting of sciences, will no doubt accompany us, like a valued friend, for life.

In this book, we shall undertake a regional study of the geology of Britain, exploring some of the extraordinary geological diversity to be found upon what is, after all, a relatively small island. In doing so, perhaps we shall come to better understand and appreciate the wonder of the natural world which surrounds us. In spite of its often rugged appearance, it is a relatively fragile world of complex interconnections and relationships, the underlying balance of which may be impacted by the burgeoning human race and the artefacts of civilisation. However, within the context of geological time, we are simply a very recent blip. Who knows what our planet, let alone Britain, will look like in a million years time? In any event, this book explores, describes and photographically documents a small corner of the planet as it exists today.

The photography undertaken throughout the research and preparation of this book is in itself perhaps worth mentioning. In this digital age it has become commonplace to artificially enhance photographs, creating larger than life images which while superficially attractive, do not represent an accurate rendition of scenes as they actually occurred. Pick up almost any glossy magazine or popular 'coffee table' book and you are treated to a barrage of artificially enhanced images which describe a world which doesn't exist. The photographs in this book, all of which were taken on-site by the author, are different. There is no artificial enhancement. The colours you see are the colours of nature. The scenes are exactly as created by nature and witnessed by the author with nothing added or removed. Other than minor corrections of exposure, the images are exactly as captured by the camera, with no filters or special effects employed. This, I believe,

Fig. 2 Weathered limestone at Deep Dale in the Peak District

is important for any imagery which may be used within a scientific context to illustrate the reality of the world under discussion. In many years to come, if someone picks up a copy of this book, I would like them to understand that what they are seeing is what was actually there at the time, not a colourful caricature. In any event, nature needs absolutely no enhancement from us. Its beauty is beyond any such human trickery.

Julian Ashbourn
Berkhamsted, UK

Fig. 3 The early morning mist lifts from the East Anglian fens

Acknowledgements

As with any book, the edition you finally hold in your hand is the product not just of the author, but of a team of individuals working behind the scenes with a common objective of producing the finished work. In this respect, I have been especially fortunate to have the guidance and co-operation of the team at Springer, with special thanks to Dr. Robert Doe and Nina Bennink. The book would not have been possible without the additional understanding and continual support of my wife Joanna, especially with regard to the long hours spent on research and photography.

Contents

In the Beginning

In the Beginning

Imagine that we are able to drift effortlessly throughout the vast universe. Our travels unveil many wonders. Bright stars dazzle us with their brilliance and smaller clusters sit like jewels upon a cloak of soft black velvet. Splashes of translucent colour occasionally present themselves and everywhere around us is beauty and majesty upon an infinite scale. But this universe of ours, in all its scale and majesty, is not a static picture. There is life, death and rebirth in a cycle of wondrous activity stretched across dimensions of time and space which we may never fully comprehend.
As we drift silently through this graceful vista, we pass through a galaxy where there is an inexorable tension building. A mass of swirling, hot gases have been caught in the vice-like grip of gravitational force, contracting them ever inwards upon themselves, increasing heat and tension again and again. The mighty gravitational force is unstoppable, the heat sears in intensity and the tension increases until the sphere of gases can no longer bear the burden of existence in their present form. As the temperature soars beyond 10 million degrees K, suddenly there is a blinding flash of light as our gaseous sphere, by now compressed into a highly massive entity, explodes defiantly into a nuclear fusion of hydrogen and helium, producing enough thermal energy to push back upon the suffocating force of gravity. A miracle of nature has occurred and a beautiful new-born star has been delivered into the universe.
The star is our own Sun. As it swirls exuberantly through the darkness, as if in the throws of a joyous fast waltz, some of the matter which was flung from its surface in that mighty moment, together with peripheral remnants of the original gaseous sphere, is caught in the Sun's own gravitational pull and is happily swirling along with it. Over an extended time period, this halo of matter slowly consolidates into a number of individual spheres which circulate the Sun in distinct orbits, according to their precise construction. Those made of hard, rocky material remain close to the warmth of the Sun while those of a more gaseous and icy construction stay a little further out. Some of them even have baby circulating spheres of their own. Indeed, a veritable family has been formed. A solar system.
This early solar system of our Sun and its planets, is a dramatic place. It is filled with swirling meteorites and other matter, crashing into the planets and each other with explosive force, sometimes adding to an individual planet's mass, sometimes blasting off splinters in collision which will themselves crash into other bodies until, very slowly, the major planets accumulate a good deal of the debris and things start to settle down a little. There are still thousands of meteor collisions, but an order has been established and relative stabilisation achieved.
Among the rocky planets closer to the Sun, lies our own beautiful Earth. Imagine being upon the Earth in those formative years of creation. What thunderous sounds and magnificent displays of light would we have experienced? Nature's orchestra in full symphonic form playing out a crescendo of dramatic sound, slowly diminishing into a gentle, steady heartbeat which would last for billions of years. For there was something very special about this particular planet. This beautifully formed sphere had its own inner life. A deep, central core of energy radiating slowly outwards and creating a protective magnetic field which would shield the Earth from the worst excesses of solar radiation while, together with other factors, would enable the formation of a mixture of gases into an atmosphere.
The almost perfect positioning of the Earth's orbit in relation to the Sun allowed it to retain a vast quantity of liquid water. A little nearer to the Sun, and this water would have vaporised,

J. Ashbourn, *Geological Landscapes of Britain*,
DOI 10.1007/978-90-481-8861-1_1,

a little farther out and it would have turned to ice. The atmosphere would work hand in glove with the Earth's oceans to eventually create unique conditions and opportunities to support that most miraculous of miracles, life itself. However, there would be much activity of a different kind upon our new planet. Activity resulting from that inner life-force, energy and heat from which would power volcanoes and move the solid surface plates around upon the Earth's mantle, enabling areas of ocean and land to perform a slow motion dance which periodically changes the very shape of continents and oceans, occasionally tearing them apart, sometimes crashing them into each other. This is the dance of plate tectonics which continues today and will be enacted so long as the Earth's heartbeat endures. Thus, we have an active atmosphere, biosphere and geosphere, all of which are cyclical and which together form our most extraordinary living planet.

The tectonic plates which form the Earth's crust, both oceanic and terrestrial are the uppermost section of the mantle. This section is itself comprised of two zones, the uppermost lithosphere and the lower, more compliant aethenosphere. The lithosphere itself is subject to constant cycles of renewal and destruction. Renewal takes place at mid-ocean spreading ridges where new magma erupts at a seam, pushing the two halves of the divided crust apart. This eruption takes place very slowly and spasmodically, causing the ocean floor to expand a small amount each year. As the Earth is of a finite circumference, for every creative boundary thus described, there has to be an equal rate of destruction somewhere. This destruction takes place at a tectonic plate boundary called a subduction zone. A subduction zone typically occurs at a plate boundary where oceanic crust meets terrestrial crust and the heavier oceanic crust slides under the lighter terrestrial crust, with the older edge of

Fig. 1 Woodlands blossom within the early Hertfordshire landscape

the oceanic crust being subsumed back into the Earth's magma layer. Consequently, the oceanic lithosphere is constantly being renewed, albeit at a very slow rate.

The process described above is something of an ultra slow motion ballet, causing the continents to pirouette, join together and then part again as they enact their timeless tectonic dance, accompanied by side effects of volcanic eruptions and earthquakes. This is the loom which creates the very fabric of our world. We join the ballet for what, in geological time frames, may be referred to as the last act, or the Phanerozoic period, starting around 542 million years ago.

In pre-Cambrian times (before 542 million years ago), there had been a huge super continent straddling the world which had subsequently broken up leaving, in early Cambrian times, a massive continent named Gondwana, which still stretched almost from pole to pole, together with some smaller remnants named Laurentia, Baltica and Siberia. Deep in the southern hemisphere, near to the south pole, a fragment off the coast of Gondwana named Avalonia, contained Wales, England, Newfoundland and Nova Scotia. By late Cambrian times, around 500 million years ago, much of the continental land mass was to be found in the southern hemisphere. Sea levels also rose during Cambrian times, as is reflected in strata of sandstone and (later) mudstone. Cambria is of course the Roman word for Wales and it is in North Wales during the early nineteenth century where much of the mapping of rocks from this period took place. It was geologist Adam Sedgwick who so named the period in the 1830s.

As we move into Ordovician times, around 480 million years ago, we see much activity as Avalonia and another fragment, Armorica, break completely away from Gondwana and move north, with a new mass of water filling in behind them named

Fig. 2 Fossil imprints in limestone from Napa Scar in Yorkshire

the Rheic Ocean. The southern part of Gondwana became heavily glaciated at this time. There is also major volcanic activity with additional land being created and appended to the continents. During this period, the mountains of the English Lake District were created in their original form from an intriguing and resilient mixture of slate, gritty sandstone, volcanic lava, ash and basalt.

Around 440 million years ago we entered the Silurian period where most of the continental land mass was still situated in the southern hemisphere, with the huge continent of Gondwana hovering around the southpole. In the northern hemisphere, the great Panthallassic Ocean was the dominant feature. However, Laurentia and Baltica were still on the move and Avalonia was closing the gap between itself and Laurentia, shrinking the Iapetus Ocean which lay between them while, south of Avalonia, the Rheic Ocean continued to expand. Armorica collided with Avalonia during this period.

As we move into Devonian times, around 400 million years ago, the topography of our beautiful planet starts to change quite dramatically. In the south, the massive Gondwana, which included much of what we now call Australia, Antarctica, India, Africa and South America, was drifting slowly towards Laurentia, which included much of what we now call North America and Northern Europe. Around 360 million years ago, Baltica, which included what we now call Scandinavia, collided with Laurentia, the impact creating substantial mountain chains. In fact, there was a good deal of mountain building going on in Devonian times, including within the areas we now call Australia and China. Armorica and Avalonia now parted company again, moving around like loose cannons upon the deck of the Earth.

Fig. 3 Life finds a way for lichen upon this exposed limestone

The Carboniferous period, around 320 million years ago, saw more drama unfold as Gondwana rotated, closed the Rheic Ocean in the west and opened up Tethys Ocean in the east as it collided with Laurentia, creating the super continent of Pangaea. Siberia was also drifting dangerously close to the edge of the super continent which had once been Baltica. Meanwhile, Armorica collided with Iberia, which included much of what we now call Portugal and Spain and the two of them collided with the southern coast of Avalonia, creating a mountain range which would stretch from modern day Germany, right through to Ireland.

Continuing into Permian times, around 270 million years ago, a significant moment in the tectonic dance is reached as Siberia finally collides with Baltica, confirming the position of Pangaea as the great super continent. It is hard to conceptualise a land mass of such proportions, but Pangaea stretched almost from pole to pole and contained most of what we would now call the modern world. However, time doesn't stand still, even for super continents, and parts of what had been Gondwana were starting to break away again.

In the Triassic period, around 240 million years ago, the Earth was still dominated by one great super continent, Pangaea and one great ocean, the Panthallassic Ocean, although the Tethys Ocean was also vast. Pangaea was moving northwards and the Gondwana part of Pangaea was still breaking up at the edges, sending fragments such as modern day Tibet, Malaysia and Southern China northwards across the Tethys Ocean. It is thought that a major mid-ocean ridge must have existed on the floor of the Panthallassic Ocean at this time, driving much of the associated tectonic activity.

Now, as we move into the Jurassic period, around 170 million years ago, things start to

Fig. 4 The tectonic ballet raises the land from the sea

look very interesting, with Pangaea starting to break up again. The landmasses of modern day North America and Eurasia are starting to evolve and Gondwana is on the move. Modern day Antarctica is actually a considerable distance from the south pole and relatively warm. But this is also a period of much volcanic activity, with major eruptions taking place in what is now Antarctica, South Africa and eastern North America, causing massive outpourings of basaltic lava.

In early Cretaceous times, around 130 million years ago, Pangaea continued to pull apart and the newly formed Atlantic Ocean was taking shape, separating the Americas from the part of Gondwana which would be Africa and Eurasia. Indeed, by later Cretaceous times, around 95 million years ago, the continents were adopting shapes which would look quite familiar to us today. But there was still much to occur. India and Madagascar had broken free from Africa and were drifting along on their own, while Australia was still tentatively attached to Antarctica.

As we leave the Cretaceous period and enter the Tertiary period, a boundary referred to as the K-T boundary, there would appear to have been much activity including a period of cooling and reduced sea level (sometimes associated with the Chicxulub meteorite strike in Mexico). There was also continued volcanism and the world must have seemed like a very troubled place. However, the tectonic dance continued apace with the Atlantic Ocean opening out, Africa moving northwards, Iberia rotating and colliding with Europe. By Mid Tertiary times, around 20 million years ago, Africa continued to approach Europe, squeezing what had been the mighty Tethys Ocean into a thin strip of sea and crumpling the lower edge of Europe, throwing up an east-west mountain chain through the Balkans. By Late Tertiary times, around 5 million years ago, the world looked essentially as it does

Fig. 5 The land and water in battle at Capel Gwynfe

today, with all the continents being where we would expect to find them, although not precisely in their present positions.

I have given a necessarily concise view of what I have called the tectonic dance, in order to illustrate the fact that land masses are subject to movement around the globe and not pinned statically in place. From this concise overview, one might form the impression that there has been a tremendous amount of hectic past activity leading to the current consolidated picture, where everything has settled down nicely. But of course, this simply isn't the case, the activity continues. The tectonic ballet will run and run. We have described a sequence of events which straddle almost 550 million years. If we were to imagine a time 100 million years from now, things would no doubt look rather different. However, we don't have to wait that long to see evidence of tectonic activity and also less dramatic, but nevertheless significant activities such as coastal erosion, which we can measure upon an annual basis and even minor movements among existing structures such as Mam Tor in the Peak District.

As we have seen, the little pieces of land which we now call the British Isles, have travelled literally around the world, variously colliding with other lands and then being pulled apart again, until finally ending up as we see them today. This astonishing history has served to create an equally astonishing little group of islands, featuring an amazing variety of rock formations and natural features crammed into a relatively small space. It is quite sobering, when observing some of these features, to think of the extraordinary history behind them and what a momentous chain of events has lead to their creation. In this book, we shall focus upon a selection of areas within mainland Britain in order to explore some of the variety and natural beauty resulting

Fig. 6 Deep layers of limestone at the heart of the Black Mountain

from that extraordinary history. No doubt we shall find some fascinating examples of habitats sitting upon limestone, sandstone, granite, basalt, mudstones and shale. Certainly, we shall find some evidence of passing ice ages and we shall try to notice some of the smaller detail within nature's rich tapestry as spread across these islands. From a geological perspective, Britain is a particularly interesting place, with visible evidence straddling a wide range of geological time. Some of the earliest rocks, from the Cambrian period and even earlier, are to be found in the northern extremes of Scotland and in various regions of north Wales. Progressing through the Ordovician and Silurian periods, we find representative rocks in southern Scotland, throughout north and mid Wales, parts of Ireland and northern England. The Devonian period is represented throughout the west country, parts of Ireland, southern Scotland and parts of north east Scotland. The Carboniferous period takes in large parts of Ireland, the English midlands, south Wales and parts of southern Scotland. The Triassic and Jurassic periods are represented throughout the English midlands and down to the south coast around Dorset. The south east of England consists of much later strata from the Cretaceous and Cenozoic periods, also visible in parts of north east England and isolated parts of northern Ireland. Thus, within a relatively small area, an impressively broad range of geological time and associated strata are represented, indicating the somewhat dramatic history of the formation of Britain.

Sitting atop this variety of bedrock is an equally wide range of habitats which support a variety of life in both the flora and fauna. Some of this is obvious and familiar and some of it less so, especially when one starts to look a little closer. Indeed, nature is overflowing with wondrous beauty at every scale. From

Fig. 7 The sea carves the coastal rocks at Little Haven

the magnificence of an evening sunset, to the tiniest insect making its way among the woodland floor. There is so much to delight the eye of the patient observer. Every little corner which one might walk by without hardly noticing, contains its own micro world of activity and associated drama being enacted daily. The closer one looks, the more one is held in awe at the complexity, ingenuity, adaptability and amazing beauty of nature's designs. One finds oneself wondering what a world without humans may have been like. What wondrous diversity would have existed? How would biological evolution have aligned itself with the tectonic dance? And what amazing richness of life would there have been in the oceans? We catch glimpses of the past through the fossil record which, while incomplete, portrays a fascinating collection of life and its wondrous diversity throughout many of the geological periods. With the help of cladistics and other techniques, we may construct various hypotheses of evolution, but we can never be absolutely sure of how things may have developed without human influence.

Early humans tended to work much more closely with nature, using its bountiful materials for protection and shelter, and being sustained by its fruits. We tend to think of their lives as being rather basic, even to the point of crudity, and yet, they consistently surprise us with their artistry and skill. Perhaps they had the best of it. There was much that they understood, and they were still able to absorb and enjoy a largely unspoiled world. As we stroll through the pages of this book, it may be interesting to wonder how many eyes have gazed upon the scenes presented here, at what point in history, and what their thoughts might have been. Of course, the scenes are not exactly the same. There will be subtle differences as habitats

Fig. 8 Life flourishes upon the woodland floor

and coastlines evolve and, of course, man leaves his mark upon the landscape. Things change. They will continue to change. But here at least, we capture an impression of the geology and associated habitats of Britain at a particular point in time. Future natural explorers and researchers may like to compare the images presented herein with their own observations.

Fig. 9 The waters slowly etch their way through the valley

Fig. 10 Revealing the colours of the Peak District sediments

An Overview of the Geology of Britain

An Overview of the Geology of Britain

The very creation of Britain as an island is a fascinating story in itself resulting from hundreds of millions of years of tectonic activity wherein land masses collided and split apart again as they moved across the surface of the globe in their tectonic ballet. Even since Precambrian times, which ended 542 million years ago, the various land masses have been pirouetting across the globe with areas which are now oceans variously covered by land and vice versa. The history of the Earth of course reaches much further back to 4,600 million years ago, but if we documented the various tectonic events just since the Precambrian period, we should have an interesting list. If just one of these events had been different, if a moving land mass had altered its course by just a few degrees, if a continental split had occurred at a slightly different position, if the timings had been slightly desynchronised, then Britain as we know it may not have been created at all. Even much later on, if ice ages, glaciations and subsequent melts had not contrived to form the North Sea and the English Channel, then there would be no Britain as a separate entity. The very existence of Britain is as a result of random tectonic and climatic activity. Consequently, Britain owes its existence as an island as much to chance as anything else. And yet, this island embodies a pocket history of much of the aforementioned activity, nicely wrapped and presented in a beautiful little package. It is indeed an island gem.

Due to the aforementioned tectonic ballet, our island gem contains remnants from land masses we now know as North America and Europe as well as oceanic and arid desert environments which were once situated in the tropics. These have all been jumbled together and fashioned by successive glaciations to produce our current distinctive land form. As a consequence, we are blessed with a rich variety of geology within a very small land mass. Similarly, a broad time period is represented, with Proterozoic rocks of between 1,500 and 3,000 million years of age sharing the island with much younger sediments from the Palaeogene less than 60 million years of age. This diversity is readily apparent as one journeys around Britain, as has been discovered during the preparation of this book. This section provides a brief overview, with the following sections providing a greater level of detail for specific areas of interest.

If we start in East Anglia and work our way clockwise around Britain we shall discover much of this interesting geological variation. On the eastern coasts of Norfolk, Suffolk and Essex, the geology is quite young with the sedimentary layers being deposited primarily within the Palaeogene period, less than 65 million years ago. This strata sweeps along the northern shores of the Thames Estuary and extends westwards well past London. It exists of shales, mudstones, sandstones, clays and of course chalk. Combinations and variations of these may be found throughout the area, with the chalk exposed at certain sites and clays and gravels at others. The coastlines tend to be relatively soft and vulnerable, especially the eastern Norfolk coast whose cliffs of soft sandstones and mudstones are easily eroded giving cause for concern, especially with the spectre of future rises in sea level to consider.

On the northernmost Norfolk coast, as one looks towards Hunstanton, we find strata from the Cretaceous period laid down between 65 and 142 million years ago. These include sandstones and mudstones and the distinctive red chalk to be found within the Hunstanton cliffs. This Cretaceous swathe extends in a south westerly direction down to the south coast where it skirts above the coastline adjacent to the Isle of Wight and briefly touches the Dorset coast. From here it also extends eastward to cover Sussex and most of Kent,

J. Ashbourn, *Geological Landscapes of Britain*,
DOI 10.1007/978-90-481-8861-1_2, © Springer Science+Business Media B.V. 2011

excluding the north Kent coastline which displays later strata. The southern part of this coastline includes the distinctive white cliffs of Dover and those of the Seven Sisters further to the west. These beautiful cliffs actually consist of many layers, often separated visually by thin seams of flint nodules, spaced fairly regularly upon the exposed cliff faces. These cliffs are also subject to erosion and one can often find sizeable chunks of the chalk which have been detached from the main cliff and are being slowly eroded by wave action. This south eastern section of the south coast nonetheless provides some interesting variation, including pebble and shingle beaches and a range of habitats atop the cliffs as one looks inland.

Moving westwards along the south coast we enter the what has become known as the Jurassic Coast, with mudstones, sandstones and limestones laid down during the Jurassic period between 142 and 205 million years ago. At Lyme Regis the beautiful limestone pavements and associated features have long attracted fossil hunters as this area has produced some of the most significant fossils from the period. Ammonites and fossil remains of small marine creatures are to be found in abundance. Unfortunately, the adjacent cliffs are somewhat unstable and the area has been prone to mudslides and partial cliff collapse. Most recently, this activity has obscured certain parts of the beach where fossils were readily found. It is almost as though nature, having once presented us with such rich finds, is now burying them once again.

On the Dorset coast, we find several well formed bays and some interesting features such as Durdle Door, where the waves have etched a beautiful archway into the limestone finger protruding out to sea. Within the adjacent bays to the east of Durdle Door, one can find similar fingers of rock stretching out to sea at low tide. The bays themselves have

Fig. 1 The chalk cliffs meet the sea at Seven Sisters

some very interesting formations of limestone and sandstone, occasionally exhibiting some distinct folding and extreme angles of stratification. Clearly there has been some dramatic underlying geological activity to form the coastline we see today. The swathe of Jurassic strata extends from the south coast in a north easterly direction to The Wash, skirting around the Cretaceous strata at the coast around Hull, before appearing again briefly in Yorkshire. In so doing, it cuts right across the lower Midlands.

As we move further westward to the Devonshire coast we find strata from the Devonian period between 354 and 417 million years of age, considerably older than the Cretaceous rocks found further east. These are mostly sandstones and mudstones which occasionally exhibit a striking visual appearance with some beautiful burnt orange and red sandstone cliffs, such as those found at Exmouth and Sidmouth, contrasting against the hues of the adjacent sea. While relatively soft in composition, these cliffs seem to be rather more durable than those further along the coast to the east. They also exhibit some distinctive layering in places which simply adds to the beauty of this coastline. An interesting exception may be found at Beer, where the coastal rocks are composed of chalk and gleam white in the morning sun. On the adjacent beaches there are many variations of pebble and sand to be found including, at Budleigh Salterton, the particularly distinctive large dense pebbles which characterise the beach at that point. These pebbles, unlike any others found on the south coast, are composed of a hard quartzite material, similar to that of the Brittany rocks which are around 440 million years of age. Speaking of pebbles, those at Chesil beach, further eastward on the Dorset coast, reach out to sea towards Portland in a

Fig. 2 The soft sandstone cliffs at Happisburgh on the Norfolk coast

long spit well over 20 kilometres in length which serves to protect the largest tidal lagoon in Britain. The Fleet lagoon forms a natural nature reserve within which many species thrive and are effectively protected by the pebble spit of Chesil beach.

There are so many interesting features along this stretch of coastline. Kimmeridge gives its name to a distinctive Jurassic clay which may be found within the bay. The clay is interspersed with limestone and there are some interesting limestone pavements and spits reaching out to sea which render Kimmeridge Bay a most interesting spot for those wishing to observe marine life in the shallows. The rocks are particularly distinctive with yellows, browns, greys and some very dark mudstones providing a colourful spectacle, interspersed by some equally interesting pebbles and overseen by the Kimmeridge cliffs with their distinctive layering of large swathes of Kimmeridge clay. The area contrasts notably with that further along to the west.

Inland from the south Devonshire coastline we find the distinctive granite intrusion which is Dartmoor. Dartmoor rises up from the surrounding area in a beautiful gently undulating plateau which is broken through in many places by the granite tors. These tors range in both size and texture, some like Haytor, becoming a favourite with rock climbers anxious to practice their skills upon these small scale 'mountains'. A little further north, Houndtor presents a subtly different picture with the granite appearing more rugged and twisted as though having taken a tortuous path before finally breaking free at the surface. Each of the Dartmoor tors has its own special character however and indeed, the granite can assume a quite different appearance at various places across the moors. Often, in the riverbeds for example, the granite may be interspersed with calcite

Fig. 3 Extreme stratification and folding of the sedimentary rocks at Durdle Door

seams with individual rocks becoming particularly colourful and distinctive. This granite underpinning provides for the ideal moorland environment, supporting the moisture upon which sphagnum mosses and a variety of flora thrive. In places, particularly to the north east of the moors, peat bogs are also to be found which can make travel on foot a little precarious. The Dartmoor granite is part of a larger granite sub-system which is also evident further along the coast at Lands End in Cornwall.

Moving further west we come to the Cornish coast with its picturesque bays and cliffs founded predominantly upon rocks from the Devonian period between 354 and 417 million years of age. These include sandstones, siltstones and a great deal of slate. Part of the Variscan geological strata which runs south to Brittany and across Europe, there is much evidence of compression and folding to be found within these rocks. In addition, there are the aforementioned granite intrusions which surface in several places, all of which serve to remind us of the dynamic processes involved in the creation of this landscape. Cornwall of course has been relatively mineral rich and mining for a variety of ores has been undertaken in the area for literally thousands of years, intensifying in the nineteenth and early twentieth centuries. The varied geology of the area has also attracted geologists with the Royal Geological Society of Cornwall established in 1814 in response to this interest. One of the areas which would have undoubtedly appealed to them is The Lizard. With its fingers of hard serpentine rock stretching out to sea it is not only the southernmost point in Britain but also a good example of an ophiolite, whereby a section of oceanic crust has been uplifted and exposed as continental crust.

As we round The Lizard and work our way around the coastline we come to Lands End,

Fig. 4 The beautiful red sandstone at Sidmouth on the south Devon coast

where the hard Cornish granite is surfaced in a spectacular and rugged manner. From here we can explore the north Cornwall coastline with its bays and sandy beaches as it reaches up to the north Devon coast with its high cliffs which often exhibit some intense folding, as may be clearly seen at Hartland Quay where the rocks adopt a concertina like form in several places. Moving north along this coastline we shall also encounter younger rocks from the Carboniferous period laid down between 290 and 354 million years ago. However, moving back inland we encounter Exmoor with its gently rolling hills which are formed upon an anticline of Devonian strata with the associated syncline dipping down northwards into the Bristol Channel. Exmoor supports both woodlands and moorlands, providing a range of natural habitats within its boundaries. Across Somerset we encounter younger rocks from the Triassic and Jurassic periods as we move up the Bristol Channel and continue our journey into South Wales.

The Brecon Beacons and surrounding area offer some interesting contrasts and show us that, as we head north, the geology is changing somewhat. Much of the area consists of Devonian sandstones, siltstones and mudstones including the distinctive 'old red sandstone' to be found across much of the area. However there are older rocks from the Silurian period, around 417–443 million years of age, and some strata from the Ordovician period from around 443–495 million years ago. The area also shows evidence of some severe faulting and folding among the rocks, with the faults extending in a northeastward direction right across the area. As we head westward we have the Gower peninsula to the south with its sandy beaches and interesting sedimentary strata and Pembrokeshire before us with its equally interesting coastline and cliffs of Carboniferous limestone. The

Fig. 5 The exposed granite of Haytor on Dartmoor

Pembrokeshire cliffs are notable for their erosional patterns and textures, including blowholes, caves, arches and other features cast into these 300 million year old rocks as a consequence of weathering and marine erosion. Indeed, when the weather turns on this coastline, the combination of rain, high winds and wave action constitutes a force to be reckoned with and its effects may be readily seen in places where active erosion is under way. The combination of Carboniferous rocks and the associated folding and compression processes have exposed seams of coal in both north and south Wales, but it is the southern coalfields in particular which played a significant part in the Industrial Revolution. In fact, the earliest reference to coalmining in the Pembrokeshire area goes back to 1324 and since this time the process of extracting coal has been steadily developed, although the mines are now mostly silent. The southern coalfield extended for nearly 90 miles across southern Wales and it is not surprising that coal mining has played such a significant part in the affected communities over the years.

Moving up into north Wales, and we find an increasingly dramatic landscape dominated somewhat by the mountains of Snowdonia but with an interesting diversity across the entire area. North Wales constitutes an interesting collection of uplifted sediments, volcanic rocks, coal seams and of course slate, which has played a significant part in the industrial history of the area. There are some ancient rocks here from the Cambrian period between 495 and 545 million years of age as well as younger Ordovician and Carboniferous strata. Indeed, the very name Cambria is derived from the Roman name for Wales. The volcanic rocks have served to shape some distinctive features on Snowdon, Cadair Idris and other peaks, providing Snowdonia with a character of its own. There are many other geological features to be found within the

Fig. 6 Waters flowing through the limestone at Aira Force in the Lake District

area, including the Harlech Dome of Cambrian strata between Snowdon and Cadair Idris and much evidence of glacial activity which has helped to shape the north Wales we are familiar with today. A journey along the Miners Path and up to the top of Snowdon unveils many of these features, including lakes and a variety of exposed sedimentary strata as well as volcanic rock on the mountain itself. To the north west of Snowdonia lies the isle of Anglesey where some of the oldest rocks in Britain may be found, dating from the late Precambrian, early Cambrian periods between 500 and 700 million years of age. In fact, Anglesey is something of a geological mixed bag as there are a variety of interesting rocks and associated strata to be found here with evidence of fairly intense folding and compression in places. The compositions of many of the sandstones and siltstones suggest that they are turbidites, originally laid down within turbulent underwater currents and consequently containing a good deal of coarse grained mixed debris. There are also quartzites and metamorphic rocks, including some which suggest an origin of undersea lava flows. The island also has limestones and a central outcrop of intrusive granite making the area particularly interesting from a geological perspective and, indeed, the island has attracted geologists since the early nineteenth century due to this interesting diversity of structures.

Across from north Wales moving inland and slightly northward we find the Peak District in Derbyshire which sits at the southern end of the Pennine hills which run like a backbone up through northern England. The southern part of the Peak District, often described as the White Peak, is limestone country with Carboniferous deposits originally laid down around 325–350 million years ago to be found, together with a good deal of quarrying activity. The northern part of the Peak District,

Fig. 7 Erratics left at High Tove in the Lake District

often described as the Dark Peak, features younger sandstones and shales overlain upon the limestone. These are the rocks exposed at the surface. However, boring activity has shown that there are lower strata of Devonian and even Ordovician origin dating back to between 350 and 495 million years ago and rocks of volcanic origin have also been found. Thus the Peak District is an interesting example of composite layers from a broad time period. Much of the landscape is moor-like with plateaus and gently rolling hills interspersed with outcrops of limestone and gritstone. There are many caves and caverns to be found within the White Peak area and minerals, especially lead, have been mined here since Roman times. Throughout the sixteenth and seventeenth centuries, millstones were also shipped throughout the country from the Peak District. It is a fascinating area of Britain, full of interesting geological features and associated history.

Continuing our journey northward and now moving back towards the west, we enter the Lake District, a wonderful example of a landscape sculpted by glacial action, beneath which lays an interesting geology. The Lake District embodies a curious combination of volcanic rock, slate, limestones, sandstones, mudstones and even granite intrusions, mixed together and garnished with, of course, a selection of beautiful lakes and tarns courtesy of the glaciations which forged through this area, scouring the softer rocks between the volcanic peaks. Some of these rocks have their origin in the Ordovician period between 443 and 495 million years ago and the combination of ancient volcanic and sedimentary rocks are to be found throughout the fells and valleys of this picturesque area. The three main groups of rocks form something of an incredibly thick sandwich with the earliest being the Skiddaw group of sedimentary rocks, overlain by the Borrowdale and Eycott

Fig. 8 Seams of calcite high up in the rocks at Arrochar in Scotland

volcanic groups which, in turn are overlain by the Windermere group of sedimentary rocks. These layers are thousands of metres in depth, creating a substantial mass of material. Around the eastern and northern periphery of this mass there are younger rocks to be found from the Devonian, Carboniferous, Permian and Triassic periods. The lakes themselves are fascinating, each with an interesting character and a story to tell as to its formation. It is an area which has inspired poets, writers and painters as well as attracting geologists from far and wide.

Moving further north through Cumbria and a patch of Jurassic rock and into the Scottish Lowlands, we come across a good deal of older strata from the Silurian and Ordovician periods between 417 and 495 million years of age. These reach across in a north-easterly direction from the west coast right up to Edinburgh, where there are also some Devonian rocks to be found as well as some small outcrops of volcanic rock. A little further north we find further diagonal swathes of volcanic and Devonian rocks reaching right across the country. There are also of course the lochs, many of them displaying a variety of rock formations and habitats around their shores, including the impressive Loch Ness which also stretches out in a north-easterly direction across much of the country. The Cairngorms provide a mountainous landscape shaped by the various glaciations of the last ice age and feature mostly Precambrian metamorphic rocks more than 500 million years of age. Now we are discovering some of the oldest rocks, not only in Britain, but which are to be seen generally. Much of this area, right into the Scottish Highlands, feature strata from the Lower Palaeozoic and Upper Proterozoic periods formed between 500 and 1,000 million years ago. In the northern highlands and outer Hebrides we find rocks from the Lower

Fig. 9 The jagged slate pushed up adjacent to Loch Long

Proterozoic and Archaen periods around 1,500–3,000 million years of age. These are very old rocks indeed. Scotland provides a mixture of igneous, sedimentary and metamorphic rock formations, often in close proximity. It is a fascinating area from a geological perspective providing many contrasts, including across the age range of visible strata. The coastal rocks often adopt a distinct appearance due to weathering and often the attachment of a variety of lichen. The lochs which open to the sea similarly exhibit a sometimes colourful appearance at low tides as the seaweed provides a rich carpet around the exposed rocks. In the mountains, a rather more rugged expression is often to be found although there is equally often a rich diversity as one moves from area to area. Scotland also exhibits some notable faults, including the Great Glen Fault which runs right across the country. It is thus a land of contrasts in many respects, yet retains a discernable overall character which attracts people from all over the world.

Heading back south into the north of England, Yorkshire in particular provides a wealth of interesting features. The Yorkshire Dales consist of a series of rolling hills, peaks, plateaus and rivers all with their own particular characteristics. This is limestone country of course, mostly laid down between 290 and 354 million years ago in the Carboniferous period. There are many limestone pavements to be found with some wonderful examples around Ingleton. Weathering of the limestone takes many forms and there are also underground streams and caverns in several areas as well as interesting above ground streams with several notable waterfalls. The Craven Dales have additional interest in the form of the three peaks of Whernside, Pen-y-ghent and Ingleborough which tend to dominate the landscape. They exhibit distinct layering, as indeed can be found throughout the Dales,

Fig. 10 The smoothly rounded landscape at Glen Lyon

often in repeating series of limestone, sandstone and shales. This often gives the landscape a stepped appearance as the different layers have various resistance to weathering. Indeed, layering and steps are a familiar theme throughout the Yorkshire Dales, especially where rivers have cut their way through the landscape. Another familiar sight are limestone scars, often accompanied with acute weathering. Napa Scar for example exhibits a veritable avalanche of broken limestone, fairly uniform in individual rock size, many of which are relatively fossil rich. The distinctive shape of many of the crags and hills owes much to glacial activity as is the case with so much of northern Britain, and it is perhaps not surprising that humans have occupied this landscape for thousands of years, with its clear waters and diverse but easily navigated habitats.

To the east, we have the Yorkshire Moors with their gently rolling hills leading down to the east Yorkshire coastline. This area consists mostly of Jurassic sediments laid down between 142 and 205 million years ago. There is an early limestone layer, interspersed with sandstones and shales, as outcropped along much of the coast, and a later limestone layer as exposed within the Tabular Hills. There are also gorges and, perhaps unexpectedly, a dyke of igneous rock injected into the central moorlands known as the Cleveland Dyke. The moors can be quite beautiful and colourful with their swathes of purple heather and golden grasses rolling across the contours of the land. While the upper moors tend to be raised and tilted up towards the north, it is a different story in the south and particularly along the south eastern coast where erosion can be significant. The coastline of the northern Yorkshire Moors is itself quite spectacular with its beautiful cliffs of Jurassic sandstone adopting a variety of interesting forms. Many particularly interesting fossils have been

Fig. 11 Weathered limestone at Napa Scar in Yorkshire

found along this coastline leading to its alternative name of the Dinosaur Coast.

As we head further south through areas of Jurassic underlying geology, we eventually return to the earlier rocks of East Anglia, laid down in the Cretaceous and Palaeogene periods between around 55 and 142 million years ago. Thus, within a short circuitous tour of Britain we have encountered a wide variety of rocks ranging from a few tens of millions of years of age to around three billion years of age. We have additionally encountered a variety of landscapes, from gentle moors, fens and plateaus to magnificent mountains, valleys and lakes. We have walked upon volcanic lavas, granites, sandstones, mudstones, shales, gritstones, chalk and almost every type of rock. We have enjoyed a variety of natural habitats and walked in the shadows of our ancestors from the Neolithic and Mesolithic periods. All of this upon a relatively small island. Such is the geology of Britain. The following sections explore these areas in greater detail and allow you to witness much of this diversity for yourself via the accompanying photographs.

Fig. 12 The mudstone and sandstone cliffs at Whitby in Yorkshire

Fig. 13 Rocks tumbling down towards Thirlmere in the Lake District

Early Settlers

Early Settlers

The British Isles have been inhabited throughout history by quite a wide variety of peoples and a complete history of settlement within the British Isles, colourful though it may be, would be outside of the scope of this book. There are simply too many developments and counter-developments as one race after another has visited these islands and influenced the culture of the day. However, it is useful to include an appreciation of the Mesolithic and Neolithic periods, from around 8,000–2,500 BC, as this time window marks a significant development in the effect that human occupation had upon the landscape.

In general terms, we think of Mesolithic settlers as hunter gatherers who lead a fairly nomadic life, following their prey as it moved from one area to another. Equally, we tend to think of Neolithic settlers as the first farmers, clearing areas of land for planting and developing more sophisticated tools. Consequently, they tended to stay in defined communities, creating effective villages. One fundamental issue we have with these generalisations concerns the transition between them. One school of thought favours a steady evolution by existing settlers from a nomadic hunter gatherer lifestyle to a more settled agricultural one. Another school of thought favours an influx of new settlers from outside Britain, bringing their agricultural ways and skills with them. The conundrum is further complicated by the relative scarcity of a decisive fossil record covering the transition period. However, this varies on an area by area basis and new evidence is being uncovered all the time.

There is much evidence of Mesolithic occupation in various parts of the British Isles, from the south-western moors, to small islands off the northern coast. Imagine the beautiful landscapes that these early settlers would have found. Rich forests with an abundance of wildlife. Natural moors and sparkling freshwater streams and rivers. A variety of mountainous and lowland areas and, of course, a rich coastline with an abundance of marine life. Following the retreat of the previous ice age, life would have blossomed across the British Isles, leaving a variety of colourful and plentiful habitats. It must have been a truly beautiful place for those early settlers, regardless of their sometimes precarious lifestyles.

And Mesolithic lifestyles in themselves were apparently quite interesting. They seemed to find their way to virtually all corners of the British Isles, taking advantage of whatever opportunities for sustenance presented themselves. They clearly roamed the coastal areas, right up to northern Scotland, becoming connoisseurs of shellfish and no doubt a variety of fish. Rudimentary harpoon like tool fragments suggest that they thought imaginatively about the eternal struggle between man and fish. They clearly were also able to travel between islands easily enough and it looks as though they would also transport specific rocks from one area to another for sundry purposes, whether purely aesthetic or functional. All of this suggests an acute awareness of the environment and how best to interact with it.

There is not a great deal of archaeological evidence to show how Mesolithic groups lived and moved across the country. However, it is clear that they used a variety of stone tools and probably prized certain types of stone above others in this context. They seem to have lived mostly in makeshift dome shaped tents covered in animal skins and also used animal skins for clothing and bags, although they probably would have had few possessions other than the essentials due to their nomadic lifestyle. However, occasional finds suggest that they did have an appreciation of beauty and perhaps carried items of personal value, almost like jewellery. This is hardly surprising. After all, why wouldn't they appreciated the beauty of

J. Ashbourn, *Geological Landscapes of Britain*,
DOI 10.1007/978-90-481-8861-1_3, © Springer Science+Business Media B.V. 2011

the world around them and delight, as we do today, in finding an unusually beautiful or otherwise rare stone and enjoy handling it and showing it to others. They roamed across England and Wales, up to Scotland and even reached the Orkney Isles, demonstrating a seafaring capability as well as curiosity. They also reached Ireland, almost certainly crossing from Scotland, and evidence of Mesolithic activity is increasingly being discovered across the British Isles. Given what must have been a relatively small total population, it is indeed interesting that these hunter gatherers managed to explore so much of the country. Equally interesting is their adaptability in moving from predominantly coastal marine environments, to inland environments of a very different nature. One can imagine family groups moving between favourite locations, perhaps on a seasonal basis, and delighting in finding new locations, as yet unknown by their colleagues. Their knowledge of wildlife, the weather and natural habitats increasing all the time and being passed down from one generation to the next, as would be their developed skills in fashioning stone and wooden tools, such as arrows, spears and axes. The distances they covered, over all types of terrain, were clearly significant given their effective coverage of the country. The boats they used to cross sometimes quite large tracts of sea, would probably have been timber framed craft covered in animal skins and sealed in some way. Such boatbuilding skills must have been painstakingly learned and developed over time. Thus, a typical Mesolithic group would have contained a variety of practical skills and detailed specific knowledge of various habitats, no doubt handed down from generation to generation. One wonders what social interactions might have taken place between family groups and whether they would have occasionally joined forces in order to roam and hunt in larger bands. It must have indeed been an interesting life.

Fig. 1 Early settlers would soon learn to use the stone to their advantage

A key factor in Mesolithic life is that the nomadic groups tended to respond to natural habitats and seasons, working with nature and enjoying its bounty without trying to fundamentally change it in any way. This was perhaps the last time that the relationship between human kind and nature would be as symbiotic, certainly as far as the British Isles are concerned. From now on, humans would be leaving a much stronger mark upon the landscape and would be seeking to bring much of nature under their control. Such activities have served to shape much of the landscape of the British Isles, changing what would have been a largely forested environment into a variety of different habitats and setting a pattern for the future.

The transition into Neolithic times and a predominantly agricultural lifestyle is an interesting one. On the one hand, one expects such transitions to be gradual with a degree of continuity between old and new. On the other hand, the evidence of such a continuity from the fossil record is not conclusive, particularly on the mainland where several sites seem to show a distinct gap in the dating of relevant artefacts. Curiously, on islands such as the Orkney Isles, the Shetlands and the Isle of Man, there is rather more evidence of a transition. Nevertheless, this distinction between the Mesolithic and Neolithic lifestyles has caused many to propose that the change was instigated by new peoples arriving from the European mainland. Peoples who had evolved skills in farming, land management and the domestication of animals. However, if this was the case, what happened to the indigenous hunter gatherers? Are we to assume that, almost overnight, they abandoned their previous lifestyle and adopted the farming practices of the new arrivals? Or is it simply that the indigenous population developed farming skills for themselves and started to systematically adopt a more settled lifestyle? It is an

Fig. 2 Standing stones such as this one on Dartmoor are to be found throughout Britain

intriguing question which remains largely unanswered to date, although future discoveries may shed more light on the matter. The conundrum is accentuated by the fact that, in many cases, early Neolithic settlement did not necessarily include a fully developed agricultural approach, but rather a continuum of transition towards the more formal agriculture of the Bronze Age. So perhaps the transition between what we understand as Mesolithic and Neolithic lifestyles was logical and gradual after all, involving both indigenous peoples and new arrivals.

In any event, the Neolithic approach was typically one of clearing land areas in order to plant crops and manage livestock, while maintaining, to some degree, the traditions of hunting and gathering. In addition, Neolithic dwellings started to adopt a greater permanence with their relatively sophisticated wooden structures and an increasing use of stone. Furthermore, they often tended to arrange dwellings in small communities, such as those found at Skara Brae in the Orkney islands and elsewhere. Consequently, Neolithic communities started to impact the landscape to a far greater degree than their Mesolithic predecessors had done, and there is a considerable amount of evidence of Neolithic life throughout the British Isles, in Somerset, Wiltshire, Devonshire, Cornwall, Cumbria, the Lake District, Derbyshire, the Isle of Man, across Scotland and of course in Ireland. These were interesting times in which the physical representation of Britain was subject to a great deal of change.

In addition to the developing understanding of agriculture, the Neolithic people were also developing a broad range of skills including the manufacture of pottery, the ability to polish and grind stone and a more sophisticated preparation of animal skins. It is probable that there was also a certain amount of trading, especially with respect to prized

Fig. 3 Our ancestors left many curious monuments, such as this one at Stonehenge

items such as carefully manufactured stone axes, from stone quarried from as far afield as Cornwall and the Lake District, and it is clear that arrow heads and arrows were being systematically developed, as were the bows used to launch them. It is also interesting to note the appearance of Neolithic trackways, often characterised by the use of hazel and birch timbers laid across a defined path packed with earth and occasionally lined with wooden stakes or stones. This suggests a regular interaction between groups, perhaps for both social and commercial purposes. The purpose of several long stone rows found around the country is rather less clear, but no doubt has some ceremonial context, as do the various stone circles. Burial chambers also played a significant part in Neolithic life and these varied from simple and small stone constructions to much larger 'barrows' or communal graves, and a variety of burial mounds, some with sophisticated entrance chambers. However, due to the proportionality of such sites found in relation to the probable population, it would seem that such ritualistic burials were probably reserved for important individuals or families.

The Neolithic peoples tended to favour hilltop locations, perhaps because it was easier to clear hilltop woodlands than those in the valleys and, no doubt, the soil drained more easily from such positions. Of course, such locations also provided a commanding view over the local terrain, affording a certain amount of security, although there is little evidence of conflict between groups. Housing became more sophisticated with indoor fires becoming a feature and animal husbandry was also developing. Curiously, in Ireland, which had no native cows, goats or sheep, these animals were successfully introduced by Neolithic farmers who must have brought them across from mainland Britain. What sort of craft they used to bring them across the

Fig. 4 Woodlands and scrubs would have been familiar territory to the early settlers

Irish Sea we can only guess at, but clearly they had a well developed capability in this respect.

There is no doubt that Neolithic settlers changed the landscape of the British Isles and this was the first time that human beings had done so. Indeed, historians and archaeologists often refer to this time as the Neolithic revolution, when the practice of farming and settled communities swept across both Europe and the British Isles. But what of the Neolithic people themselves? What were their family associations, their beliefs, their ambitions and how did they feel about the world around them? Today, some would describe them as a primitive culture, yet they were clearly intelligent and, in many ways innovative. Their curiosity around new ideas and places to be explored lead them to some extraordinary advances and achievements. They additionally started to develop art and had an eye for aesthetics. How they must have appreciated the natural beauty of the lands around them.

There is a little spot on Dartmoor which I discovered by chance. Down in the valley beyond Black Tor, invisible from the higher ground, exists an idyllic little area where three tiny streams converge and roll down the valley to become the source of the river Meavy. Here there is a sheltered natural indentation with a little waterfall which has a special peace and beauty of its own. On the edge of this indentation is evidence of a small Neolithic stone enclosure and, close to the convergence of the little streams, a tiny hut circle. It is charming to think that a Neolithic family also came upon this little piece of paradise and appreciated its natural beauty enough to establish a little camp, bearing in mind that, almost certainly, this would not have been their primary dwelling. I sat upon the same rocks that they would have done and my gaze fell upon the same beauty that would

Fig. 5 Neolithic stone circles are to be found all over Dartmoor

have so delighted them. I drank from the same clear waters as they did and looked back up the hillside, as they must have done, as the morning sun illuminated the golden grass and the ever present Dartmoor clouds sailed low across the sky, as they do to this day, ready to create rainbows at a moments notice. I at once felt a close and warm kinship with them. They were, after all, my ancestors. They saw with their eyes, loved with their hearts and worked with their hands, in an environment which was beautiful.

Fig. 6 The stone rows retain many mysteries as to their original function

Fig. 7 Early settlers were no strangers to the coast and seas around Britain

The South Coast

The South Coast

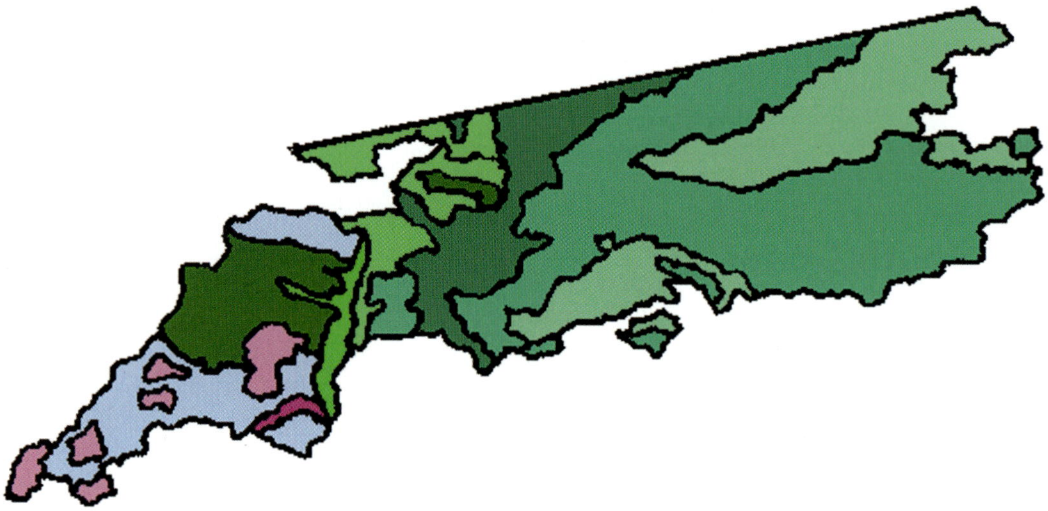

Period - Age (MYA)	Key
Quaternary 0-65	
Cretaceous 65-142	
Jurassic 142-205	
Triassic 205-248	
Permian 248-290	
Carboniferous 290-354	
Devonian 354-417	
Silurian 417-443	
Ordovician 443-495	

Neoproterozoic 545-1000	
Cambrian 495-545	
Metamorphic Rocks	
Palaeozoic 500-1000	
Proterozoic 1500-3000	
Igneous Rocks	
Intrusive	
Volcanic	

The South Coast

Summary of Geology

The geology of the south coast of Britain is varied with Devonian rocks approaching 400 million years of age in the far west, giving way to younger Triassic and Jurassic rocks of around 200 million years of age as we move eastward towards Dorset, with a huge swathe of Cretaceous rock between 65 and 140 million years of age in the south east, interrupted around the Southampton area by a band of comparatively young Miocene rock of between 2 and 24 million years of age. In the far west we find granite outcrops such as at Lands End, with a variety of sedimentary layers as we move eastward, varying in colour from beautiful red hues to more traditional greys and, in the south east, of course, the chalk, with significant bodies on the coast at Seven Sisters and Dover. Moving slightly in land, there is a swathe of very young Miocene rock around the London and home counties area, moving across diagonally to the Suffolk and Norfolk coasts. The south coast of Britain is thus quite varied in both the relative age and appearance of its cliffs and visible coast line. From gleaming white chalk to the burnt amber and reds of western sandstones and various shades of grey. The beaches also show an interesting variation, from the large dense pebbles at Budleigh Salterton through shingle beaches in Dorset and the pebble beaches at Hastings, interspersed with a variety of sands.

Overview

The south coast of Britain, or to be more correct England, from Dover in the east to Lands End in the west displays an interesting mix of geology and habitats representing a broad historical period. It features a colourful mix of chalk, sandstone, limestone, shingle and pebble beaches, bays and estuaries within a relatively short distance of around 350 miles or so. For an explanation of this diversity, we need to travel back to the Cretaceous period, around 140–65 million years ago, when there was much geologic activity as the Atlantic ocean widened and the area we now call the British Isles had a much warmer, more humid climate, with lush swamps and lagoons supporting a variety of life. Around the middle of this period, within the surrounding warm waters, billions of algae bloomed and their skeletons would sink to the seafloor, layer upon layer, creating the chalk which we now see at various points along the south coast. Around this time there was significant and deep geological activity which effectively pushed up the western rocks, tilting them slightly towards the east. Evidence of this activity may be seen along the Jurassic coast, an area which encompasses the Dorset and east Devonshire coastlines and features an interesting variety of rock formations and colours as the older rocks visible in the west are superseded by younger layers. The Cretaceous was of course also the time of the dinosaurs and a wide variety of interesting flora and fauna, until the mass extinction at the end of this period changed things forever. The Jurassic coast is famous for its fossils which reflect these interesting times. Indeed, fossil hunters have been visiting the area for hundreds of years and much has been learned through the activities and collections of enthusiastic amateurs as well as more formal explorations. This fossil record allows us, in many cases, to corroborate our theories of evolution and sometimes leads us towards important new discoveries as the south coast reveals its many secrets of times past. Unfortunately, some of the areas most noted for fossils, such as the cliffs and beaches around Lyme Regis, are subject to coastal erosion and landslides which are slowly but

J. Ashbourn, *Geological Landscapes of Britain*,
DOI 10.1007/978-90-481-8861-1_4, © Springer Science+Business Media B.V. 2011

inexorably changing the topography of the area. Such natural occurrences may reveal new fossils, or may simply obscure the area most favoured by fossil hunters. No doubt natural historians will be watching such developments with interest.

The first sight many have of England are the white cliffs of Dover, as these are the closest point to the European mainland. They were the first sight that greeted Julius Caesar during the Roman invasion of Britain in 55 BC and he mentioned them in his own account of this campaign. Indeed, Dover and the white cliffs have featured in many literary works, including those by Charles Dickens, Daniel Defoe, William Shakespeare and Samuel Pepys. They were also popularised in the World War II song by Vera Lynn (the 'forces sweetheart'), becoming a symbol of peace and hope for a better future in Europe. The white cliffs have certainly captured the imagination of humans over the millennia and it is easy to understand why as you gaze upon their fragile beauty and the contrasting green of the countryside they support. The white cliffs at Dover, formed in the Cretaceous period, may be loosely considered in terms of lower, middle and upper layers sitting upon a base of dark green sand. The lower layer often appears a little smoother and more compressed while the more nodular middle and top sections tend to contain a good deal of flint. In some places, such as in the vicinity of the Langdon cliffs which overlook the Straits of Dover, the cliff top is not uniformly clearly cut, but has in places a stepped area at the top, exposing a small secondary area of chalk, containing many flint nodules. The downland atop these cliffs represents a rare habitat which supports an interesting variety of flora, some species of which are thought to pre-date the last ice age and have become well adapted to the chalk soil of the region. Interestingly, this soil layer is remarkably

Fig. 1 The chalk cliffs of the Seven Sisters on the Sussex coast

thin, as can be seen on the upper cliffs where, in places, it is just a few centimetres thick and, of course, the chalk cliffs themselves are very fragile with a good deal of erosion taking place every year. Nevertheless, rare and beautiful plants such as the sainfoin, pyramidal orchid and birdsfoot trefoil may be found, together with a variety of grasses, providing some interesting textures and colours to contrast with the chalk. Birds also make good use of the white cliffs, with species such as fulmars and peregrin falcons nesting within the cliff face and one can always find a passing seagull. The white cliffs of Dover are not merely symbolic, but encapsulate in their peaceful elegance an important geological period, while supporting an interesting and quite rare habitat. As such, they provide a good introduction to the south coast of England.

As we move westwards, past the marshlands at Romney and the sandstone cliffs beyond Fairlight, we encounter historic towns such as Hastings where, on October 14th 1066, William, Duke of Normandy, initiated the Norman conquest during the famous battle and where the sand and pebble beaches have attracted visitors ever since. A little further west, we come to some beautiful white cliffs along the Sussex coastline known as the Seven Sisters, so named in recognition of the seven discernable peaks that constitute their profile. Like the cliffs at Dover, the Seven Sisters are chalk cliffs which originated during the Cretaceous period. From the beach, one can discern a distinct stratification of these cliffs, with periodic injections of narrow flint seams among the compressed chalk. It is interesting to note the relative regularity of this stratification and to consider the timescales involved in its creation. From a single spot, one can gaze at a beautiful picture of millions of years of creation. The relentless action of the sea causes erosion at the base of the cliffs, resulting

Fig. 2 The characteristic dark sand at the foot of the white cliffs of Dover

eventually in significant collapse of the chalk. The evidence of this process can be seen among the numerous chalk boulders which litter the narrow beach, some of them actually sitting in the sea and becoming smoothly rounded by the action of the waves. The more recent of them appearing chalky white, while their predecessors show signs of being graced with lichen, mosses or seaweed. Even the sea itself is slightly coloured by the presence of these boulders, with a milky translucent quality where the leading waves meet the pebble beach.

Above the Seven Sisters lay the chalk grasslands of the south downs, which support a variety of interesting flora and fauna. This relatively fragile habitat is maintained as a result of hundreds of years of managed grazing, ensuring that invasive scrubland and coarser grasses do not intrude. Consequently, delicate wild flowers such as the milkwort and round-headed rampion are able to thrive, creating a micro-world of interesting colours and textures. In addition, the environment provides a home for many interesting insects including a variety of beetles and colourful butterflies such as the Adonis blue and, as with the white cliffs of Dover, a variety of birds are to be found, some nesting within the cliffs themselves, others elsewhere on the downs. There are also adjacent areas of salt-marsh, providing a haven for wading birds and supporting their own special collection of marsh plants which have adapted to such an environment. Overall, the Seven Sisters and surrounding area provide a fascinating mix of geology and natural habitats, the latter albeit influenced somewhat by mankind. For those with an appreciative eye, there is much beauty to behold within this broad but delicate environment.

As we move further west, we approach an area which, for geologists, is perhaps the most

Fig. 3 Fallen segments of cliff at Seven Sisters

interesting of all, the coastline between west Dorset and east Devon, often referred to as the Jurassic coast, as the rocks effectively record marine conditions during this period, around 200–140 million years ago. Prior to this period, during the Triassic (250–200 million years ago) England was thought to be encapsulated within the huge Pangaea super continent before it broke apart to form what became the current continental picture. Consequently, this area of the south coast would have most probably been a hot, desert land adjacent to a western mountain chain, and this is reflected in the western-most area of the Jurassic coast around Exmouth and Sidmouth in Devon, where dramatically beautiful cliffs of red rock may be found, contrasting with the greyer limestones found both further east and west of this point. Interestingly, these red rocks are interrupted adjacent to the village of Beer, where an intrusion of chalk cliffs appear. It is thought that these cliffs escaped the expected erosion in this area due to them being faulted downwards to the level of the older Jurassic rocks. Whatever their history, they certainly provide an interesting contrast along the coastline. Another oddity lies to the east in the form of Chesil beach, a 29 km long barrier of graduated pebbles which protects the largest tidal lagoon in Britain. This barrier is thought to be the result of long shore drift activity, predominantly during and immediately following the last ice age, where pebbles would have been pushed along the coastline in an easterly direction, aided by rising sea levels. In any event, the result is a naturally protected area which supports both fresh and salt water species, with extensive meadows of eelgrass and a diverse mixture of seaweed, sponges and anemones. In addition, the natural lagoon supports a significant number of water birds for whom the sheltered location is precious. It is an interesting phenomenon which illustrates how beach

Fig. 4 The edge of a limestone pavement at Lyme Regis

structure, itself influenced by oceanic processes, can in turn influence the terrestrial land beyond.

Many coastlines are of course subject to erosion of one sort or another and the south west coast of England has an additional problem in the form of mudslides, rendering many cliff areas unstable, particularly between Lyme Regis and Charmouth. It was in this area, at Black Ven, where one of the largest coastal mudslides in Europe occurred in 1958 and again in 1959, spilling right out onto the beach. This is prone to happen where sandstones overlay the Jurassic clay and the cliff bases are subject to continual erosion and the process will doubtless continue into the future, possibly accentuated if we experience a rise in sea level. An earlier landslide at Bindon in 1839 became something of a tourist attraction. It was actually more of a 'blockslide' as a large section of land pulled away from the main coastline leaving a chasm in its wake. Eventually, the area was turned into a national nature reserve and has become an important, if somewhat unstable wilderness habitat with dense scrub, woodland and stretches of open land supporting an interesting variety of flora and fauna. The geology of the area includes mudstones from the Triassic and Jurassic periods, overlaid with later shales and sandstones from the Cretaceous, topped by later Cretaceous limestones. A wealth of fossils have been found in the area, providing a further insight into these interesting times. However, the erosion continues and, at the time of writing, further landslides have occurred in the area, covering much of the limestone beach with softer mudstones, broken into random boulders amid slides which are particularly hazardous, especially at high tides where the sea meets the edge of the mud.

A little further east at Lulworth Cove, a natural bay exists due to continual erosion of the

Fig. 5 The eastern side of Durdle Door in Dorset

limestones and clays at a site where a river ran into the sea. The erosion slowed as it hit the massive chalk deposits, creating the almost perfect horse-shoe cove that exists today. However, it is likely that continued erosion will continue to change this part of the coastline. Adjacent to the cove, the limestone rocks display some interesting stratification, with near vertical and heavily warped layers such as those seen at Stair Hole, reminding us of the intense activity which lead to the formation of this south western corner of England. Immediately west of Lulworth Cove is Man'O'War Cove and Durdle Door, where we find some interesting transitions of limestone and chalk, together with the effects of erosion by the relentless tides. At some places, the stratification of the rock is near vertical and often twisted, while elsewhere it is uniformly horizontal, suggesting some interesting historic activity. Indeed, between the small pebble beach and the rugged habitat on top of the cliffs, a variety of textures and colours may be found, sometimes quite surprising in their contrasts. Further west at Sidmouth, another contrast is provided by the dramatic red sandstone cliffs of the Triassic period which stretch from here to Exmouth and exhibit some distinct layering. Their colour suggests that these rocks originally formed on land under hot, arid conditions such as those found in desert areas and have subsequently weathered, with traces of iron minerals producing the distinctive rusty red colouring. These beautiful sandstone cliffs appear surprisingly rugged along their lower visible strata which, in places, receives a direct onslaught from the waves. They also exhibit very clear sedimentary layers of varying height, telling their own story of the creation of this part of the coast.

At various points along this coastline, one can find a variety of interesting pebbles, many of them originating at Budleigh Salterton to the

Fig. 6 Waves lapping the base of the steep sandstone cliffs at Sidmouth

west of Sidmouth. The interesting thing about these pebbles is that they are composed of an extremely hard quartzite, similar to the 400 million year old rocks found in Brittany, but unlike any other rocks found in the south of England. They exhibit an interesting range of hues, some of them a light bluish grey, others a pale yellow and some patterned with calcite seams, all of them relatively well smoothed by wave action over the millennia. They are mostly quite large in scale, with many of them measuring 15 cm across and few of them less than 7 or 8 cm across, providing an interesting beach landscape. It is thought that the Budleigh Salterton pebbles may have originally formed in one of the giant rivers which flowed into Triassic desert areas around 240 million years ago and finally ended up embedded in the local cliffs, from where they have been detaching themselves for thousands of years. Their relative hardness ensures their resilience as they are transported along the coast, from west to east, by wave action. They are to be found in varying sizes at various points along the coast, from Chesil beach, where the graduation in pebble size is quite marked from larger in the west to smaller in the east, to Hastings on the south eastern coast where they are altogether smaller and rather more uniform.

The variety of both geology and natural habitats along the Jurassic coastline reflects both an interesting past and nature's ability to adapt to changing situations. It is an extremely beautiful and diverse area, packed into a small section of the coast of a relatively small island. Yet this small stretch of coastline is home to a colourful geology and associated habitats, with the reds, greys, browns, yellows and white of the various rocks complemented by a variety of grasses and an array of beautiful but delicate wild flowers. Along this coastline, a variety of wild birds and other creatures complete this extraordinary picture of nature's

Fig. 7 The contrasting soft sandstone and hard pebbles of Budleigh Salterton

diversity. However, it is a relatively fragile picture, subject to continual change, both as a result of natural processes and the burgeoning expansion of mankind.

As we move westwards from this area we encounter the coastlines of western Devon and Cornwall, which offer yet more variety. The underlying geology of this area is part of the Variscan belt of crustal deformation, formed when tectonic plates collide, in this instance during the formation of the Pangaea super continent. The Variscan belt is in evidence in the mountains of Portugal and Spain to the south and in Germany, the Czech Republic and Poland to the east. It is also evident in south western Ireland, south Wales and Devon and Cornwall. The resulting geology often features distinctive folded strata of slate, sandstone and outcrops of granite, such as that found on the higher ground on Dartmoor in Devon and Bodmin Moor in Cornwall. This gives the Cornish geology in particular a distinctive character and the resultant variety of minerals has generated much interest over the years, including a good deal of mining activity. Furthermore, it has provided a coastline of rugged beauty as we approach the western extremes of the south coast.

An area of particular interest along this coastline is the Lizard peninsular which, apart from being England's most southerly point, is a rare example of an ophiolite, a section of oceanic crust which has been uplifted and exposed as part of the adjoining continental crust. This phenomenon is quite rare, although another example may be found in the Shetland Isles. The Lizard is consequently composed primarily of serpentine, a rock which has undergone a great deal of metamorphic transformation, whereby low silica rocks from the ocean floor are oxidised and hydrolysed with water, creating the distinctive and very dense dark green rocks to be seen reaching out like fingers

Fig. 8 An unexpected outcrop of chalk at Beer

from The Lizard. This part of the coast is exposed to the Atlantic ocean and the full forces of the weather, including occasional gales of some strength. Coupled to the rugged coastline and adjacent areas of partly submerged rocks, it is little wonder that the history of Cornwall is punctuated with tales of shipwrecks and smuggling activities, providing the inspiration for many novels. Nevertheless, the Lizard peninsular is an outstandingly beautiful area supporting a surprising variety of colourful flora, due in part to its southerly latitude. On the western side of the peninsula, Kynance Cove exhibits a rugged and beautiful vista of hard rock formations which are themselves a relic of the Variscan deformations. Moving northward from this point, the rocks start to change in character, becoming more slaty and less secure as we move past Porthleven where an interesting sandbar has isolated a pool of the sea, creating an inland lake. Either side of this sandbar the rocks are subtly different from those at the tip of the peninsula, but still exhibit a history of somewhat dramatic activity with clear signs of folding and deformation. Continuing around the peninsula and following the coast in a north westerly direction, we eventually come to Lands End, the most westerly point of England, similarly rugged in its geology, featuring large areas of exposed and weathered granite. The area has historically been subject to various mineral mining activities, no doubt influencing the local habitat which supports an interesting array of mosses and lichen among its flora. Moving northwards we come to Cape Cornwall, where the rugged coastline is complemented in places with an interesting variety of pebbles and boulders, as if a transition is taking place between the southern and northern Cornish coast lines. Indeed, as we head further towards northern Devon, the exposed rocks may adopt a slightly different appearance in texture but are no less dramatic

Fig. 9 An exposed layer of Kimmeridge clay atop the mudstone in the bay

in form, often with clear signs of deformation betraying their violent origin during the Variscan collisions.

In summary, the south coast of England is an area of extraordinary diversity with a particularly rich geological and natural history. Within this small stretch of coastline one may observe many millions of years of the Earth's wondrous development and encounter a variety of habitats and associated wildlife, which seem determined to cling on to this changeable environment. There is evidence to be found of several distinct geological periods and events which seem almost to unfold before us as we travel along a relatively small distance. Indeed, one may easily travel from one end of this coast to the other within the course of a single day. But perhaps the most extraordinary aspect of all, is the sheer beauty and majesty which unfolds at almost every juncture of this extraordinary coastline.

Fig. 10 An exposed island of sandstone at Budleigh Salterton

Fig. 11 Rugged sandstone cliffs at Seaton

Dartmoor

Dartmoor

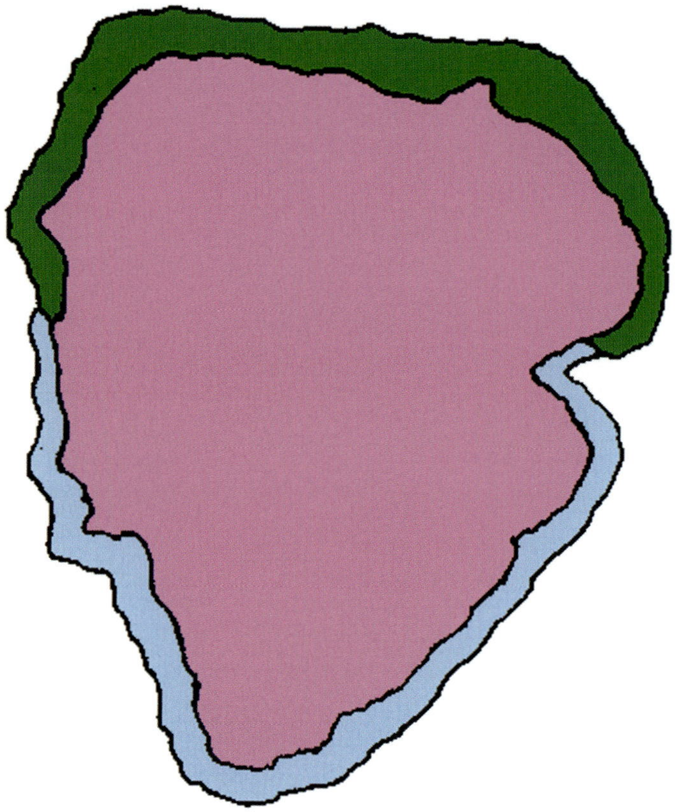

Period - Age (MYA)	*Key*
Quaternary 0-65	
Cretaceous 65-142	
Jurassic 142-205	
Triassic 205-248	
Permian 248-290	
Carboniferous 290-354	
Devonian 354-417	
Silurian 417-443	
Ordovician 443-495	

Neoproterozoic 545-1000	
Cambrian 495-545	

Metamorphic Rocks	
Palaeozoic 500-1000	
Proterozoic 1500-3000	

Igneous Rocks	
Intrusive	
Volcanic	

Dartmoor

Summary of Geology

Dartmoor, which rises up to 619 metres above sea level and occupies an area of around 625 sq. km, is founded primarily upon intrusive igneous rock. This granite intrusion is part of a massive batholith which extends in mounds from Dartmoor, right across to the Scilly Isles, just west of the coast of Cornwall. Dartmoor however, is the largest of these west country granite regions, which were formed in the late Carboniferous/early Permian period around 280 million years ago when the granite intruded into the series of Devonian and Carboniferous shales, sandstones and limestones. The enormous heat and pressure of this intrusion created a series of contact metamorphic rocks around the periphery of the granite, followed by mineral seams as the rock cooled and hydrothermal activity took place. In this section of the book, we shall also explore the interaction between humans and the underlying land form which is Dartmoor.

Overview

The south western peninsular of England, features a number of moorland areas, notably Bodmin Moor, Exmoor and Dartmoor. They are all interesting in their own right, but Dartmoor in particular is very special for a number of reasons. Its archaeology and history are fascinating enough, as are its natural habitats, but there is something more. Dartmoor has a special magic all of its own. An intangible draw which has been attracting humans for around 6,000 years or so. Is it simply the landscape, or the sense of peace and tranquillity to be found on the moors? Perhaps, and yet other locations have these qualities and can boast far more spectacular landscapes. Its hard to provide objective reasoning as to why Dartmoor exerts this special pull, but for some, the attraction is as strong as ever and calls them back to the moors again and again.

While much of the surrounding area consists of Devonian and Carboniferous sedimentary deposits, Dartmoor sits predominantly upon granite which would have been originally formed around 280 million years ago as the granite was forced upwards from the molten rock layer beneath. As the new, hot granite cooled and contracted, joints and fissures were created, allowing the injection of water and minerals which, together with its particular composition, gave the Dartmoor granite its unique properties. Atop this granite layer would have been a layer of softer sandstone or mudstone which would have systematically been worn away. Around 60–30 million years ago, what we now call the British Isles lay closer to the equator and consequently featured a much hotter and more humid climate, causing weathering of the granite due to acid attacks from wet, rotting flora. Then, around two million years ago, came the start of the last ice age, wherein the cold temperatures and expansion of freezing water caused further destruction and erosion of the granite. The ensuing repeated cycles of glaciation and the associated effects of freezing and thawing caused much weathering and erosion, leaving the more exposed tors which we can see today. The resulting Dartmoor granite is both highly distinctive and, sometimes, exquisitely beautiful. Furthermore, variations may be found at different locations. Sometimes the expected feldspar rich conglomerate, sometimes a quartz dominated mix with distinctive crystallisation, sometimes a darker and smoother composite and, occasionally, a layered composition with striking contrasts. The smaller granite rocks to be found in stream beds are often particularly beautiful. Indeed almost everywhere you go on Dartmoor, you are greeted by extrusions of granite of one sort or another. Various types are in evidence

J. Ashbourn, *Geological Landscapes of Britain*,
DOI 10.1007/978-90-481-8861-1_5, © Springer Science+Business Media B.V. 2011

including the Tor or Giant Granite, often quite dark in colour and of fairly course grain size, the Quarry Granite, also of dark colour and coarse grain size, Pink Granite, lighter in colour, often speckled with pink grey and cream and fairly coarse grained, and Blue Granite, of similar grain but speckled with blue grey and cream.

The granite found on Dartmoor, introduced during the late Carboniferous period, is actually reflected throughout Devon and Cornwall and beyond to the Scilly Isles. It is part of a larger batholith, a huge intrusion of granite within the Earth's crust which, as a result of the erosion of other rocks, becomes exposed at strategic places. During this intrusion into the host rocks, mostly shales, sandstones and limestones, a variety of metamorphic rocks were created and, as the granite cooled, hydrothermal activity lead to the creation of various mineral rich seams within both the granite and the surrounding host rocks, setting the scene for what would become a significant mining activity on the moors. Tin and copper were particularly important in this context, but there were also ores of arsenic, lead and iron with traces of zinc, tungsten and cobalt also to be found. Hydrothermal activity also caused a certain amount of kaolinisation of the granite, leading to the clay deposits to be found in south west Dartmoor and elsewhere in the region. Kaolinisation is a process whereby the feldspar minerals within the granite are decomposed as a result of the internal circulation of high temperature water, forming deposits of the white clay, kaolin. And, as everyone knows courtesy of popular fiction, there are peat bogs which cover around a third of the area of Dartmoor. Given its granite base, the above average rainfall and a predominance of sphagnum moss and other moisture retaining flora, it is perhaps not surprising that the high moors can be wet underfoot, even during the

Fig. 1 Water running down from the moors at Black Tor

brightest weather. Add a layer of peat, varying from a thickness of a few centimetres to a few metres, and it is equally unsurprising to find the blanket bogs upon the plateaus. Several rivers, such as the Dart, Tavy and Walkham in the north and the Plym, Yealm, Erme and Avon in the south drain the area. The River Teign drains the eastern edge of the moors to the south and, on the western side, the River Lyd makes a spectacular exit from Dartmoor through the Lydford Gorge, a classic example of the interaction between water and rock as the Lyd carves its way through the landscape.

Although Dartmoor lies slightly south of the glacial limits of the four ice ages of the Pleistocene period, it was nevertheless affected by these events. In particular, the freeze-thaw cycles caused a good deal of weathering of the granite, splitting huge blocks on the higher ground and littering the slopes and valleys with boulders, which we now refer to as clitter. Further weathering and erosion from acidic waters has left us with the distinctive landscape of protruding granite tors which characterise Dartmoor. They are not all granite based however. There is an exception. Brent Tor in the western moors is composed of basaltic lava which would have flowed around 350 million years ago into the shallow sea which covered the area at that time. Interestingly, Brent Tor is particularly rugged in character, perhaps as a result of continued weathering over the millennia, and does not display the smooth basaltic appearance which one might expect. Brent Tor is immediately recognisable from a distance, courtesy of the little church which sits perched right on the top of this massive basaltic outcrop.

Of course, Dartmoor would have looked a little different after the ice age. The retreating ice gave rise to a fertile area which, as the climate became warmer, was quickly populated by woodland. The analysis of ancient pollen

Fig. 2 Looking down upon the Burrator reservoir

shows that much of this woodland was oak, particularly on the lower slopes and valleys, with mixed deciduous trees higher up and, on the highest ground, fewer trees and more open moorland. However, the overall composition of Dartmoor in mesolithic times would have undoubtedly been a lush and beautiful woodland. Within these woodlands would also have existed a variety of fauna as nature flourished and populated the area. Among the larger animals would have been deer, pig and possibly oxen, providing scope for the hunting and gathering lifestyle of prehistoric peoples as they wandered across the moors. However, the availability of such woodland animals would have been much lesser in quantity to those found on the more tundra like plains. Perhaps this may have been a factor in the transition between mesolithic and neolithic life styles. Nevertheless, mesolithic peoples were obviously attracted to Dartmoor and it seems likely that they were the first to clear small areas of woodland between around 5,500 and 4,300 BC, creating grazing areas for animals where they could be more easily controlled and, when the time came, killed. The combination of repeated clearing, burning and grazing would have served to create the peat bogs, referred to earlier, which subsequently covered much of the high moors. Discovered fragments of worked flint point to a mesolithic human presence in many parts of Dartmoor, often in the proximity of streams and springs where they may have established temporary camps.

The transition to the neolithic period and a more settled lifestyle is evident on the moors where pollen records indicate the clearance of much larger trees, suggesting the use of more sophisticated tools, and the planting of various herbs and grasses. More permanent structures start to appear, including various types of burial mounds and tombs, some of them showing a sophisticated use of stone, which

Fig. 3 The water cutting its way through the rock at Lydford Gorge

must have been the result of a more communal way of working. Indeed, even some of the simple stone 'hut' circles would have involved a degree of collaboration among individuals in order to select, transport and establish the various stones, some of which would have been very heavy. Simultaneously, other community members would no doubt have been fashioning axes, preparing animal skins, tending crops and undertaking various everyday tasks according to their relative position within the group and associated skills. There is also evidence of territorial 'centres' being developed, perhaps for simple community use, or perhaps in order to provide a general administration for a given area. In any event, it is clear that the neolithic settlers on Dartmoor were becoming increasingly well organised.

As the neolithic period advanced, so did the building work of the early settlers. The simple burial mounds gave way to more sophisticated box like tombs and stone cairns, although cairns may also have been associated with non funerary functions. Stone rows also appeared on Dartmoor and their significance is still not well understood, although burial cairns and cists are often found in proximity to these rows, more of which are to be found on Dartmoor than anywhere else in Britain. Various stone circles were also erected by neolithic Britains, including of course Stonehenge. However, many much smaller stone circles appear on the moors, as do a number of single standing stones called menhirs, the significance of which is also not well understood although, presumably, they served some ceremonial purpose.

As the neolithic moved into the bronze age, the farming lifestyle became even more prevalent and field systems were established, together with more sophisticated dwellings. The bronze age houses tended to consist of circular stone walls or 'circles' probably with a timber framed thatched roof of some

Fig. 4 Twisted and weathered granite exposed at Hound Tor

kind. Furthermore, such structures were not only for human occupation but also served as grain stores and shelter for domesticated animals. In parallel, tools become more sophisticated and the use of bronze was introduced, with tin and copper being readily available in this south western corner of England, due to previous geological activity. As the bronze age gave way to the iron age, further developments in agriculture and the use of tools enabled settlers to become better established and, towards the end of this period, hill forts started to appear, settlements with surrounding ramparts, suggesting that these settlers were becoming more territorial. Curiously, the high moors tended to be gradually evacuated at this time, with more activity on the lower slopes and valleys. During medieval times, farming became well organised with an apportionment of land between farmers. However, the changeable climate and relatively acidic soil lead to variable results and occasionally areas were abandoned, only to be re-farmed at a later date. In general, the area has not been exposed to the sort of continual intensive agriculture that has changed so much of the British Isles. Consequently, Dartmoor retains a certain amount of its natural habitat and charm. The advent of the wool industry in the fifteenth and sixteenth centuries brought a more consistent prosperity and this is echoed in the building of several churches as well as more organised settlements, eventually leading to the Dartmoor we know today.

The natural habitats of Dartmoor are interesting and varied with the bogs on the higher moors surrounded by heath and grasslands with broadleaf woodlands in the valleys, creating a diverse and beautiful landscape. However, due to its relative position, the higher moors in particular are subject to above average rainfall and high winds, constraining the variety of wildlife that can exist happily in such an environment. Nevertheless,

Fig. 5 A beautiful and distinctive example of Dartmoor granite

certain species have adapted to these conditions and one can find a variety of moorland birds, notably the ground nesting birds such as the skylark, wheatear, meadow pipit and others, although the numbers of breeding pairs are thought to be in decline and this is naturally a cause for concern, an exception being the stonechat which seems to be actually increasing in numbers. Walking through the moors and grasslands, one is occasionally surprised and delighted to come across a skylark or wheatear fluttering up from the grass or gently chattering away in the background. The Dartmoor ponies are an attraction for many visitors to the moors and may loosely be divided into two categories. The more common and less shy ponies are of medium size with well proportioned legs and not unlike ponies to be found throughout the British Isles. These ponies will happily mingle with human visitors and will not be adverse to taking titbits from them. The less common ponies are the smaller, more rugged moorland ponies which tend to have much shorter legs and long coats. These are rarely found close to the roads but may be encountered out on the moors. They tend to be rather shy and will often avoid close proximity with humans.

In addition to the wildlife on the moors there exist a variety of flora, including grasses, rushes, heathers, various wetland plants such as the heath spotted orchid and ivy-leaved bellflower and occasional hay meadows with their own variety of plants. Lichen are also to be found on the moors, as are mosses, often clinging tenaciously to the rocks in exposed areas or more comfortably among the woodlands. Given the sometimes severe climatic conditions and relatively acidic soil of much of Dartmoor, it is interesting to find such a luxurious carpet underfoot. A carpet which can display a beautiful array of colours and contrasts from soft golden grasses to the purples, yellows and reds of the heath, sometimes

Fig. 6 Calcite crystals embedded in the granite are a common sight on Dartmoor

alternating as you traverse an area. Interspersed with the occasional natural pool or sparkling stream and, of course, the ever-present and magnificent granite in all its variety, Dartmoor is indeed a most unique and beautiful place. It is also steeped in history and of great importance from both an archaeological and geological perspective. But above all, Dartmoor is captivating and communicates at a deep, fundamental level, causing one to muse upon how diverse and beautiful the British Isles must have been prior to significant human occupation.

Dartmoor thus provides an interesting example of how a geological feature has shaped the existence of life within its vicinity. The varied granite is interesting in itself, but the hydrological and geological processes around it have enabled a series of habitats which have subsequently been exploited by natural flora and fauna and, of course, humankind. It is a small, yet important part of Britain.

Fig. 7 The granite tors occasionally adopt some strange forms such as here at Staple Tor

Fig. 8 The rugged but weathered granite of Haytor

Fig. 9 The distinctive profile of the main outcrop at Haytor in typical Dartmoor weather

North Devon

North Devon

Period - Age (MYA)	*Key*
Quaternary 0-65	
Cretaceous 65-142	
Jurassic 142-205	
Triassic 205-248	
Permian 248-290	
Carboniferous 290-354	
Devonian 354-417	
Silurian 417-443	
Ordovician 443-495	

Neoproterozoic 545-1000	
Cambrian 495-545	

Metamorphic Rocks	
Palaeozoic 500-1000	
Proterozoic 1500-3000	

Igneous Rocks	
Intrusive	
Volcanic	

North Devon

Summary of Geology

Northern Devon is founded upon sedimentary rocks originally laid down during the Devonian and Carboniferous periods between 290 and 400 million years ago. However, the region was subject to enormous pressures during the Caledonian Oregeny, a period of tectonic plate collision and mountain building, causing the sedimentary gritstones, sandstones and shales to be buckled and folded into some extreme forms, often with near vertical stratigraphy. Consequently, the geology of North Devon is characterised by an east-west anticline and syncline whose crest occurs close to the coast of Exmoor and within which exist a series of smaller folds. The Exmoor plateau rises to around 517 m above sea level and features gently rolling hills with lush vegetation. The coastline features a variety of rock formations, including the highest cliffs in England at around 400 m high, and some of the oldest rocks in the Valley of Rocks area. The rocks themselves vary in density and colouring according to the exposed sedimentary layers, and sometimes exhibit a slaty quality with distinct lines of cleavage. Minerals exist throughout the region, including a vast quantity of haematite ore, running in east-west lodes.

Overview

Northern Devon and the associated coastline exhibits some interesting and diverse geological features, brought about by an equally interesting history. Much of the underlying rock is from the Devonian period around 416–359 million years ago (Devon giving the period its name, due to the ancient rocks found here) with further strata from the Carboniferous period around 359–299 million years ago and later. However, such stratification is not uniform or straightforward, due to massive folding and compression of the rocks around 300 million years ago as two tectonic plates collided, forcing the rocks up into a small mountain range which has subsequently been eroded. This tremendous force has twisted and buckled the rocks into some remarkable forms which may be observed at various points along the coastline, Hartland Quay providing a very good example, where much of the exposed rock is compressed, concertina fashion, into a pattern of opposing zigzag strata. Other rocks jut out from the beach at an assortment of angles, displaying an historic stratification in fine and clear detail. Similarly, at other places along the coast, such stratification is easily observed.

Touching the coastline and straddling the Somerset and Devon borders lies Exmoor, an interesting plateau of gently rolling hills founded upon sedimentary rock layers which display an interesting transition between the ancient Devonian and Carboniferous eras. Much of this sedimentary deposition is around 200 million years or more of age, making it one of the oldest natural features in Europe. One wonders at the huge variety of creatures who existed at the time of each layer of stratification, what their lives were like and what they saw around them at each point in time. What secrets of life are locked deep within these beautiful rocks? There is little fossil evidence as the bulk of these rocks originally formed in desert like areas where life was no doubt sparsely distributed. The rocks consist mostly of siltstones, mudstones, grits and sandstones, with quite rugged exposed surfaces and occasionally a slate like cleavage. Indeed, in places the rocks appear rather fragile with evidence of a good deal of surface erosion, yet equally they remain massive and imposing in some places. It is thought that the high plateaus of this area escaped glaciation during the last ice age, probably acting as the southern boundary of

J. Ashbourn, *Geological Landscapes of Britain*,
DOI 10.1007/978-90-481-8861-1_6, © Springer Science+Business Media B.V. 2011

the glacier's reach. There remains some controversy however as to the southernmost point of glaciation, especially as some of the boulders found in coves on the north Devon coast do not seem to match the adjacent cliffs, as if they had been placed there randomly by some giant unseen hand, or simply dragged and left by retreating glaciers. However the area, glaciated or otherwise, would not have escaped the extremes of weather and temperatures associated with the ice age and such factors, including variations in sea level, would have been instrumental in shaping the topography of the cliffs and landforms we see today. Interesting features abound including evidence of a submarine forest at Porlock, various knolls and ridges along the coastline and a variety of cobble sizes across different beach areas.

The earliest layers of rocks are the Lynton slates which represent the exposed core of an east-west anticline, sloping gently eastwards from Lynton. The Valley of the Rocks at Lynton displays some of the earliest Devonian rocks in beautiful and sometimes dramatic formations, with clear stratification as the high cliffs reach down to the sea. The valley itself is a dry valley, exhibiting typical periglacial features associated with the harsh climate of the last ice age. The exposed coastal rocks clearly show the effects of weathering, yet there is an abundance of life clinging to these rocks including a variety of colourful lichen and mosses. The topsoil supports an equal variety of rugged shrubs and bracken, making the area rich in flora and supporting various insects and other forms of wildlife. This is truly a most interesting area, a jewel like cameo of nature at work. Immediately to the east, at Lynmouth, where the river Lyn kisses the sea, there are many quite large boulders strewn across the landscape, evidence of the effects of spasmodic flooding over many thousands of

Fig. 1 The gently rolling hills of Exmoor

years whereby the heavy waters rolling off of the higher moors have dislodged and carried boulders along their path, depositing them along the river beds and out into the delta. They provide yet another interesting contrast within the area.

To the west, we find younger sandstones, shales, slates and even limestones as we move along the coastline, some of them surprising us in their particular formation, with much folding in evidence. These sedimentary layers would have originally been formed in warm shallow seas, but there are boulders to be found also, and a variety of pebbles, suggesting a good deal of subsequent activity, possibly as a result of glaciation originating further north in Scotland. This view is corroborated by the raised platforms along the cliffs on the way to Saunton, cut by wave action at a time when the sea level was much higher. In the Taw-Torridge estuary there are an impressive series of sand dunes to be found, providing yet more diversity along this interesting coastline as well as a haven for wild life.

The north western peninsula at Westward Ho features a series of limestone and slaty rocks which reach out to sea like a series of rugged pointing fingers, cradling a variety of dark pebbles between them, some of them quite large. Further along the coastline, such pebbles give way to a more colourful shingle ridge, but here they seem dark and foreboding. At other points, rocks burst out of the shoreline in near vertical stratification, thrust up by some tremendous force. The adjacent cliffs are equally dramatic with another raised beach platform cut many metres higher than the current beach, upon which rugged rocks jut upwards towards the sky in extreme angles of stratification. It is an impressive sight from any perspective and no less so when one gets closer and examines the various strata and fine detail exposed at every level.

Fig. 2 Complex stratification in the Valley of Rocks

Equally rugged grasses adorn the sometimes sandy soils which sit on top of these cliffs, providing yet more visual contrast and an important platform for wildlife. Interestingly, the eastern side of Westward Ho contains a rare submerged forest, embedded in peat deposits overlying the bedrock, providing more evidence of the higher sea levels which completely swamped this coastal forest around 6,000 years ago. So much geological history and activity expressed within a small section of coastline makes this a truly interesting area.

Moving in a south westerly direction along the coastline towards Cornwall presents an equally interesting picture, culminating in the dramatic stratification to be found at Hartland Quay, where the rocks have been pushed up and compressed into some very distinctive patterns. In some places, a series of alternating V and inverted V shaped layers are clearly discernable, in others the rock is simply bent back upon itself. Elsewhere, jagged dark rocks of almost vertical stratification burst out of the sea. These are sandstones and mudstones from the Carboniferous period, around 320 million years of age, laid down originally in a warm, shallow and probably brackish sea. At the beach level, dark mudstones prevail and look a little foreboding as if still reeling from the violent action which pushed them up at such extreme angles. The cliffs themselves display an interesting stratification of many layers, providing some colourful contrasting bands of muted oranges, dark greys and browns. In places along the shoreline, rugged outcrops burst out of the shore at exaggerated angles, as if they were the far corner of a much larger slab of rock. Some of these are also quite colourful, with their surfaces relatively smoothed by wave action and weathering. A little higher up the cliff face, similar outcrops appear more rugged, often hosting a

Fig. 3 The high and rugged cliffs of the North Devon coast

good degree of flora, clinging on to the rock face tenaciously in the face of some occasionally severe weather.

The whole of the north Devon coastline is interesting in one way or another and it certainly displays some interesting and often spectacular geology with signs of both dramatic early formation and subsequent weathering. It is thought that glaciation during the last ice age reached this area from the north, resulting in a periglacial climate with long periods of freezing and subsequent thawing causing a good deal of weathering and erosion. This is reflected in some of the local rock structures. The transition between the Devonian and Carboniferous periods is captured in much of the geology of Exmoor and north Devon, with the darker, often fossil rich mudstones and limestones of the Devonian giving way to lighter sandstones or, occasionally, alternating layers of the two. This pattern reflects some fairly dramatic sedimentary action, possibly triggered by earth tremors or violent storms, releasing turbid flows of sand into suspension, which would subsequently settle into new sedimentary layers. At more peaceful times, mudstone layers would develop at their usual rate, interleaving with the lighter sandstones as described. This pattern is repeated throughout much of north Devon, with some variations which suggest that some beds were laid down in slightly brackish water, as might be found in a protected lagoon or estuary basin, and some in full marine conditions, a situation reflected in the fossil content within the beds.

In summary, northern Devon hosts a wealth of fascinating geology, much of it reflecting ancient dramas of tectonic earth movements and climatic change. It is also extremely picturesque, from the gently rolling lush green hills of Exmoor, to the rugged terrain of the Valley of Rocks and surrounding areas. As we move westwards towards Cornwall, the coastline reveals some amazing and spectacular scenes, often

Fig. 4 Boulders upon the rocky beach at Lynton

quite beautiful in their structure, colouring and inherent contrasts. A veritable 'secret garden' of rich discoveries contained within a small section of the English west country. To the south, we have Dartmoor and the south Devon coast which constitutes part of the Jurassic Coast and is equally interesting with a variety of textures and colours. Indeed, Devon as a whole is a most interesting part of Britain from a geological perspective, providing a range of rock formations, granite intrusions and sediments all within a quite compact area.

Fig. 5 The exposed rocks are sometimes of a loose and fragile composition

Fig. 6 A natural cove provides an interesting relief along the coastline

South Wales

South Wales

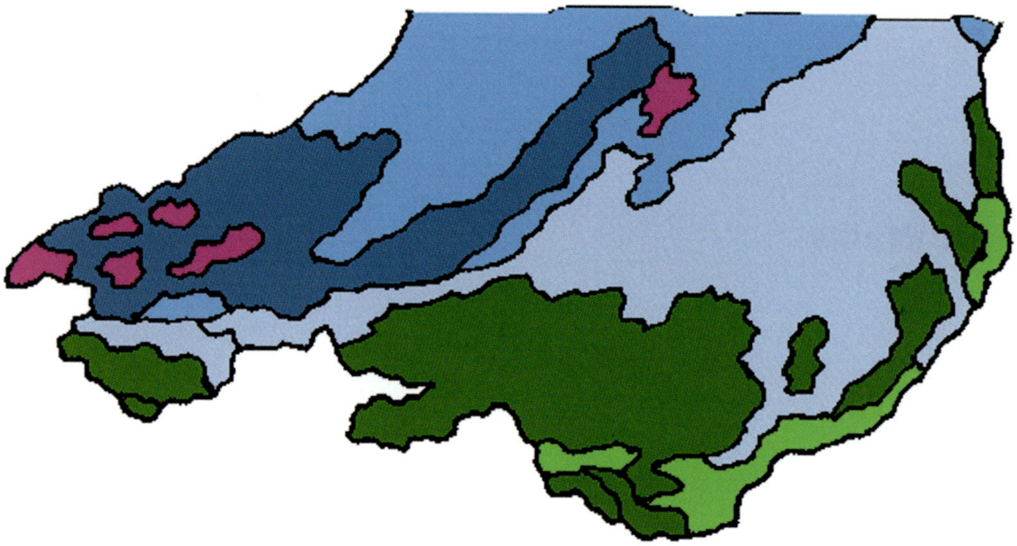

Period - Age (MYA)	*Key*
Quaternary 0-65	
Cretaceous 65-142	
Jurassic 142-205	
Triassic 205-248	
Permian 248-290	
Carboniferous 290-354	
Devonian 354-417	
Silurian 417-443	
Ordovician 443-495	

Neoproterozoic 545-1000	
Cambrian 495-545	

Metamorphic Rocks	
Palaeozoic 500-1000	
Proterozoic 1500-3000	

Igneous Rocks	
Intrusive	
Volcanic	

South Wales

Summary of Geology

The geology of South Wales represents a broad range of time and associated strata. There are Ordovician rocks of between 495 and 443 million years of age along the northern edge of the main peninsula in the vicinity of Cardigan to New Quay and then stretching across into Wales in a south west, north east swathe towards Llandrindod Wells. Silurian rocks of between 443 and 417 million years of age are present in the north west of South Wales, reaching up in a broad swathe to North Wales. Large parts of the south east where South Wales touches England feature Devonian rocks of between 417 and 354 million years of age, which are also present on the western tip of the peninsula around Pembroke. Further east along the coast towards Llanelli and Swansea and reaching up from the Bristol Channel towards the Brecon Beacons National Park are younger Carboniferous rocks of around 354–290 million years of age and there is a small arc of Triassic strata of around 248–205 million years of age reaching across from Pyle towards Cardiff. On the coastline itself, from around Porthcawl to Penarth there is evidence of Jurassic rock formations of around 205–142 million years of age. Thus, a history of over 350 million years is reflected in the geology of this interesting south western corner of Britain.

Overview

South Wales has an interesting mixture of coastlines, rolling hills and mountain areas featuring a variety of natural environments and underpinned by an interesting and varied geology. Even relatively localized areas can display a variety of geological forms due to the turbulent history of the area being periodically covered by sea or subjected to compression pressures causing faults and folds among the strata. Fforest Fawr, within the Brecon Beacons national park for example, has quite a wide ranging geological underpinning. The oldest rocks, sediments from the Ordovician period, exist in a swathe between Llangadog and Llandovery in the north west, while later Silurian rocks lay to their south around the area of Mynydd Myddfai. A broad band of Devonian red sandstones stretch almost across the whole area, recording more arid conditions and, in the south, a band of Carboniferous limestone reflects conditions of shallow tropical seas, atop of which exists a coarser layer of Millstone Grit, probably carried down from the north by rivers. There are even coal seams to be found within the area, completing an interesting picture of geological diversity within a relatively contained space. Furthermore, there are numerous faults, mostly within the Ordovician and Silurian strata, with three main faults, or disturbances, cutting into and across Fforest Fawr. The Tawe Valley Disturbance originates from Swansea in the south, runs up the Tawe Valley through Ystradgynlais and reaches right into the centre of Fforest Fawr. The Neath Disturbance runs on a north by north easterly course along the Vale of Neath from Pontneddfechan towards Crickhowell. The Carreg Cennen Disturbance in the north of the area also follows a north easterly course, past Carreg Cennen Castle and on towards Sennybridge and beyond. A contrasting set of lesser faults among the Carboniferous rocks run in a NNW to SSE line. It is perhaps not surprising to find such features as the formation of these older rocks coincided with much tectonic activity as Avalonia was finding its way among collisions which thrust up mountain ranges during a period referred to as the Caledonian Orogeny, whose eroded remnants may be found among the mountains of North Wales, the Lake District and Scotland. The original mountain range would no doubt have been of magnificent proportions, the erosion of which enabled

J. Ashbourn, *Geological Landscapes of Britain*,
DOI 10.1007/978-90-481-8861-1_7, © Springer Science+Business Media B.V. 2011

great rivers to carry debris southwards during Devonian times into the area we now call South Wales. This explains the significant swathe of Devonian red sandstone to be found within the Fforest Fawr area which, itself was subsequently uplifted and subject to further erosion.

The Brecon Beacons themselves lay south of the town of Brecon and feature some of the highest peaks to be found in the south of Britain with Pen Y Fan at 886 m, Corn Du at 873 m, Cribyn at 795 m and Fan Y Big at 719 m. The summits form a horse-shoe shaped ridge and are constructed predominantly of the Devonian red sandstone referred to earlier. They are perhaps the focal point of the Brecon Beacons National Park, although there are many interesting landscapes surrounding them. There is ongoing debate around just how far south the last glaciation, around 12,500–11,500 years ago, actually reached, although most would hold that it did reach down to the area of the Brecon Beacons, if not well beyond. Certainly, around the base of the mountains, there are features reminiscent of steep sided moraines which would corroborate the view that at least small glaciers were active in this area.

Moving south westwards, the Gower peninsula is itself an interesting area, sitting on the edge of a syncline which dips down to the north of Swansea where the late Carboniferous layers at its core constitute the coalfields for which the area is known. The rocks exposed around Swansea and the Gower tend to be older as they represent the higher edges of the syncline where there is evidence of much folding due to compression forces exerted around 280 million years ago as land masses collided. This folding extends not only throughout the Gower peninsula, but southwards across the Bristol Channel to the North Devon cliffs where there are some dramatic examples to be seen. The primary line of folding is along an east-west axis, suggesting that pressure was being exerted from the

Fig. 1 Near vertical stratification of the limestone on Black Mountain

south, crumpling the coastal terrain as though it were the edge of a piece of paper. Along with the folding, the Gower exposes several faults where the rocks have been folded or compressed to the point of fracture, often along the fold lines, further corroborating the enormous pressure to which they were subjected during this phase of tectonic activity. Along the Gower coastline there are some wonderful examples of rocks thrust up through the sandy ground at a variety of angles, sometimes exposing many sedimentary layers within a single feature. Just north of the coast, the intensity of folding eases into a much broader pattern as we enter the large syncline and the coalfields.

The cores of the primary anticlines of the Gower expose strata of the late Devonian period around 350 million years of age. These are the oldest rocks to be found in the area, overlain by later sediments when the Gower was awash with broad rivers carrying sediments down from the mountainous north towards the sea which at the time would have covered the area we now call Devon. The fairly intense erosion of these northern mountains caused the sediments to be quite coarse in places, and this is reflected within the strata of coarse sandstones and even pebbly conglomerates to be found within the strata. As sea levels rose during the early Carboniferous period, the area was further swamped with marine shales and sediments being overlaid upon these earlier coarse layers and, as things settled down with less sediments being transported from the north and the warm waters clearing, a limestone layer was formed which, in places, is up to 800 m thick but thins progressively as we move north across the Gower and in-land. As the waters rose, the lush forestation within the river deltas was buried, to be re-born as new growths during periods of low sea level, only to be buried again under shales and sediments during higher sea levels as the cycle continued. In such a fashion, the coalfields of South Wales,

Fig. 2 The rugged shoreline at St. Annes

which played a significant part in the industrial revolution, were systematically created.

Moving westwards to the Pembroke coast we find an equally interesting expression of Welsh geology. There is much folding and faulting in evidence among the predominant Carboniferous limestone, in areas along the southern edge of the Pembroke peninsula, courtesy of the Pembroke syncline running along an east-west axis. South of Tenby, just below Penally, Giltar Point exposes the limestone as it dips towards the south east. From here, one can look across the water, and indeed across the axis of the syncline, to Caldey Island where, at Drinkin Bay there are some dramatic formations among the limestones and sandstones, with near vertical stratification in places. Moving westwards along the coast from Giltar Point, Lydstep point protrudes into the sea with some equally dramatic formations among the limestone shales with a wide variance of grain size, from smooth layers to quite rough pebbly structures, indicating a loose assemblage of sediment, as well as extreme angles of stratification and evidence of a significant fault. Continuing westwards to Stackpole Quay, another fault runs along a north-south axis, separating limestones on the eastern side from mudstones to the west. The rocks are again folded and presented in places at extreme angles as we continue west to Barafundle Bay, a picturesque spot with yet more dramatic stratification among the sometimes fossil rich limestones. The southernmost cliffs west of St. Govan's Head display an intricate pattern of marine sculpture, cut into the limestone as the sea has exploited faults and weaknesses in the rock via a process of continual erosion. The steep, rugged looking cliffs display some distinct features including the Green Bridge of Wales, a natural arch cut into the rock with a span of more than 20 m. Indeed, this entire stretch

Fig. 3 Rocks trundled down through the valley at Capel Gwynfe

of coastline displays a wealth of interesting features among the limestone due to erosion effects, including several small coves and caves. Moving around the nose of the Pembroke peninsula to Angle Bay with its own series of small coves, there is much evidence of folding among the limestone with variations between fine and coarse grain layers with the coarser shales dominant at either side of the bay and a thrust line running through the coves where the limestone shales are thrust up over the main limestone. The story is similar along the Pembroke coastline, with distinct folding and erosion creating a dramatic coastal landscape.

No discussion on the geology of South Wales would be complete without mention of the coalfields and the important part they have played in the economic and social history of the area. The coalfields stretch right across South Wales, from St. Bride's Bay on the western coast, all the way to Pontypool in the east in an elongated basin of Carboniferous rocks. There are coalfields of course in North Wales, but it is the southern coalfields which are larger. The dramatic folding and faulting referred to earlier has meant that some of these coal seams were pushed up and hence made all the more accessible. Such a bounty was sure to be exploited and naturally this has been the case, with reference to coal mining in Pembrokeshire as early as the year 1324. Initially, this would have been something of a part time activity for the farming community, no doubt primarily for their own consumption although the possibility of using coal in commerce must have quickly occurred. Early mining methods were simply a matter of finding a relatively exposed seam and digging the coal out of the hillsides, propping up the resultant tunnels and caverns with timber. There would always have been an element of danger due to collapse and slides among the disturbed landscape and early miners would

Fig. 4 Sedimentary layers pushed up out of the sands on the Gower

have developed an understanding of this reality, enabling them to conceive ways of preventing such occurrences. Thus, the mining industry was born and the rolling hills and valleys of South Wales would never be quite the same again. Neither would the community as, by the sixteenth century, coal mining was becoming an industry in its own right, requiring a dedicated labour force and the necessary infrastructure with which to exploit this natural harvest. An associated supply industry of trolleys and haulage was consequently developing in order to transport the coal, as was expertise around tools and extraction methods and associated industries such as ironworks were also developing.

As the industry progressed throughout the seventeenth and eighteenth centuries, deep shafts, sometimes hundreds of metres deep, begun to be cut into the rich seams underground, necessitating lifting mechanisms in order to transport both men and coal to and from the coal face. These mechanisms developed into the cages and winding houses, with their huge wheels reaching high into the skyline, which came to epitomise coal mining for many individuals and communities. By the turn of the twentieth century the mining process had become much more mechanised with automated cutters and conveyors to rapidly remove the coal from the face to the waiting trolleys. What was once a curious supplemental activity for farmers was now a huge industry in South Wales which, by 1913 had no less than 620 mines employing 232,000 men, producing more than 50 million tons of coal. By 1975, this had shrunk to 42 mines employing 30,800 men to produce around 8.5 million tons of coal and now, the last deep mine in operation at Hirwaun, on the southern edge of the Brecon Beacons national park has finally closed, its bounty exhausted. This must be particularly poignant for the miners co-operative who pooled their resources to re-open the

Fig. 5 A variety of sediments and angles at Barafundle Bay

mine in 1995 following its closure the previous year.

This boom-bust cycle produced fortunes for many industrialists while, for the miners themselves, the toil was often hard and dirty. In the early days, the coal was cut by hand and accommodation for the workforce and their families sparse and often cramped. Conditions improved slowly, but it was not until the late 1930s that pithead baths were introduced, enabling the miners to shower and change clothes before returning home. Before that time, they would have returned to their families, tired and with the coal dust clinging to every surface. It is interesting that some of these same men would find a different sort of expression within the many Welsh choirs that sprung up in such areas, creating something beautiful from such dark surroundings. Perhaps this land was touching them in more ways than one. And of course, throughout all of this time, there was always an element of danger associated with working deep underground, with the occasional disaster reinforcing this realisation for many mining communities.

Throughout the latter half of the twentieth century, the mines were closed, one after another, causing a significant unemployment issue for communities who had themselves developed in response to, and around the industry of coal mining. It represents an interesting example of communities interwoven with the fabric of the land and their position within the economic structure. Nowadays, most of the mines have gone and the many hundreds of thousands of men who worked them over the years have become memories of times passed. But they were special times in many ways and, while tourism seeks to compensate as a major industry for the area, there are still echoes of the coalfields of South Wales to be heard, for those with an inclination to hear them.

Fig. 6 Inverse folding of the sandy mudstones at Freshwater East

Fig. 7 Weathering produces some interesting hues at Little Haven

Fig. 8 Diverse sedimentary layers uplifted at Whitesand

Fig. 9 Beautiful limestone layers lifted up to the sky on the Gower

North Wales

North Wales

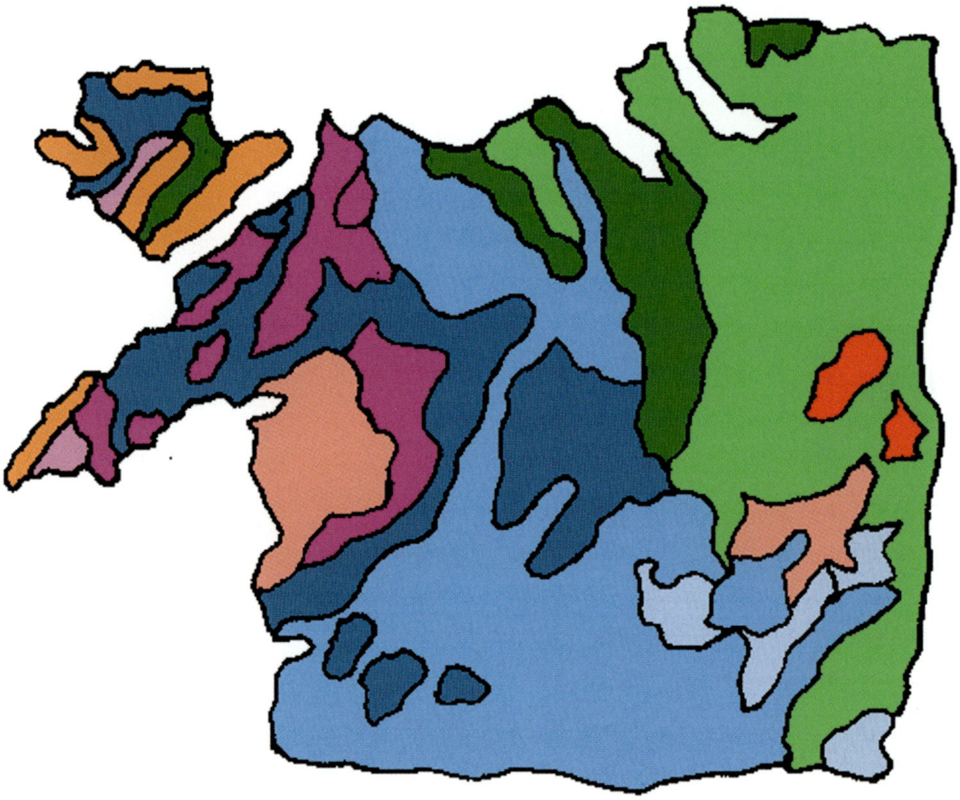

Period - Age (MYA)	*Key*
Quaternary 0-65	
Cretaceous 65-142	
Jurassic 142-205	
Triassic 205-248	
Permian 248-290	
Carboniferous 290-354	
Devonian 354-417	
Silurian 417-443	
Ordovician 443-495	

Neoproterozoic 545-1000	
Cambrian 495-545	
Metamorphic Rocks	
Palaeozoic 500-1000	
Proterozoic 1500-3000	
Igneous Rocks	
Intrusive	
Volcanic	

North Wales

Summary of Geology

The geology of North Wales spans a broad time period from late Precambrian rocks on the island of Anglesey more than 500 million years of age, to much younger traces of Carboniferous sediments less than 300 million years of age found in the north east. There are traces of volcanic rock to be found in the north west and igneous seams on Anglesey and parts of the north western peninsula, indicating an interesting history. Furthermore, parts of the landscape are shaped by more recent activity during the last ice age when glaciers scoured out the valleys around Llanberis, Nant Gwynant , Tal-y-Lyn and elsewhere. In addition to Cambrian rocks of between 495 and 545 million years of age found in the north west, there are swathes of Ordovician origin, with rocks of between 443 and 495 million years of age, reaching down from Anglesey in a south east direction, with earlier Silurian rocks, around 417–443 million years of age, completing much of the picture. Later Devonian rocks of between 354 and 417 million years of age, while occurring mostly in South Wales, are occasionally to be found in the south eastern extremities of North Wales. Overall, it is an area rich in geological variety and with an interesting history which, in certain areas, is unequivocally bound to the underlying rock. It also provides some of the most beautiful landscapes to be found in Britain, with a variety of small mountains, lakes, rolling hills and valleys.

Overview

When contemplating the landscapes of North Wales one is, of course, immediately attracted towards the Snowdonia region and Snowdon herself (Yr Wyddfa) who rises 1,085 m to stand as the highest peak in Wales. Certainly the area is rich in beauty and natural splendour and, like many such areas, has an equally interesting geology underpinning this rugged magnificence. While the familiar valleys and mountains vista we see in Snowdonia today was sculpted primarily by the last ice age, the origin of the underlying rocks lies much further back in time when, indeed, Wales did not exist as such. Tectonic movements throughout the Cambrian, Ordovician and Silurian periods caused many collisions as the super-continents broke apart and new land masses drifted across the surface of the Earth. One such land mass, Avalonia, destined to become the foundation for much of Wales, continued to collide with its neighbours throughout the Devonian and later periods until finally, in the Cretaceous it was itself torn between what we now know as North America and Europe. Snowdon and the surrounding area are formed primarily of Ordovician rock, much of it of volcanic origin. From here down to Cadair Idris further south, extends the Harlech Dome which is formed mostly of Cambrian gritstones and mudstones, although Cadair Idris itself is largely of Ordovician volcanic origin. These sedimentary layers were established prior to the volcanic activity which shaped many of the mountains in Snowdonia and subsequently impacted upon the sediments themselves, transforming them into the slates and gritstones with which we are currently familiar. Glacial activity around 18,000 years ago scoured out the broad U- shaped valleys, such as Llanberis and Nant Gwynant, adding the finishing touches to this characteristic Welsh landscape.

The Ordovician strata, around 495–443 million years of age, consist mostly of mudstones and siltstones with occasional volcanic areas, caused by compacted volcanic ash, and injected igneous seams. Consequently, there are often fossils to be found where these layers are exposed, reminding us of the marine origins

J. Ashbourn, *Geological Landscapes of Britain*,
DOI 10.1007/978-90-481-8861-1_8,

of these sediments. Strata from the later Silurian period, around 443–417 million years of age, are similarly composed primarily of mudstones and siltstones. The rocks from these periods, which make up a gootd deal of North Wales, were subsequently compressed, folded and faulted by mountain building events during the early Devonian period, around 417–354 million years ago, creating metamorphic conditions which, in turn, gave rise to the formation of large areas of slate, a rock which became an important natural resource for the area, finding its way on to the roofs of countless Victorian homes as well as being used for other construction purposes. Indeed, slate has played a significant part in the history of North Wales and this is echoed at the National Slate Museum at Llanberis, the Llechwedd slate mine at Blaenau Ffestiniog and elsewhere, with outcrops of slate to be found throughout the area.

Rocks from the later Devonian period, around 417–354 million years ago, while present in the south and east of Wales, do not surface in North Wales. However, it is thought that Devonian sediments once covered most of Wales and have subsequently been eroded away over time. Later Triassic strata, originally laid down in desert environments around 248–205 million years ago, were deposited during a time of intense rifting which created the distinctive Vale of Clwyd towards the north east and there are small Carboniferous deposits also in the north east. However, the bulk of North Wales is founded upon rocks of Ordovician and Silurian origin, underlain in places by Cambrian or even Pre-Cambrian strata, as in isolated areas of the Llyn peninsula and on Anglesey. Nevertheless, there exists a variety of interesting textures to be found within areas such as Snowdonia for example, where time has exerted its influence, not only in the metamorphic activity previously referred to, but in general weathering where one can see the effects of erosion

Fig. 1 The river Conwy carving its way through the rocks toward the Conwy falls

and freeze-thaw cycles upon the exposed rocks. The Llanberis pass is particularly interesting in this respect, displaying a rugged landscape with some quite large boulders to be found strewn across, often in a seemingly random fashion. Snowdon herself, while of volcanic origin, hosts a variety of interesting smaller scale textures and formations as one traverses from the lower slopes towards the summit, including the occasional calcite seam contrasting against the darker grey sediments and some colourful displays of lichen upon the exposed surfaces. This historic volcanic activity was the foundation for many mountainous features in North Wales and also the catalyst for the creation of rich mineral seams which have subsequently been exploited, from the bronze age onwards. Copper in particular has been mined around Snowdon, Rhobell Fawr (close to Dolgellau) and in other areas, where such activities contributed to the development of sulphide ore separation methodologies which have endured to the present day. There are even veins of gold to be found around Dolgellau, as well as sulphides of iron, lead, zinc and other minerals. It is an interesting area in this respect.

To the north west lies the island of Anglesey, separated from the mainland by the Menai Straight, a hazardous stretch of water around 23 km in length and a little over 3 km at its widest point, but reducing to around 200 m at the narrow region where it is bridged. Differential tides at either end of the straights cause some quite strong currents to flow through this channel with some non-obvious reversals occurring, making it a potentially dangerous area (albeit supporting a particularly interesting benthic environment). This dramatic juxtaposition with the mainland is perhaps in keeping with Anglesey, whose own geological history would appear to have been quite dramatic. There are a number different groups of rocks making up the island, the

Fig. 2 Extreme weathering of the rocks at Llanberis

older ones reflecting turbidity currents and debris flow as well as metamorphic activity, suggesting proximity with some ancient destructive plate margin. The turbidites consist of sandstones and siltstones, often in alternating layers, with a variety of grain size according to location, appearing in several of the named groups (notably the South Stack Group, New Harbour Group and the Gwna Group). Anglesey has been of interest to geologists for hundreds of years and the various formations are thus well documented and frequently discussed as to their origins and stratigraphy, as so much is still uncertain. There are also conglomerates with sandstone and basaltic pebbles suggesting a volcanic source and, in central Anglesey, a granite intrusion into surrounding high grade gneisses is to be found, probably associated with ancient island arc magmatism of some sort. In the south east, a linear belt known as the Blueschists contain metamorphic rocks which have been subject to unusually high pressures, such as might be found within a subduction zone.

These dramatic rock formations reflect a history of undersea land-slides, volcanic activity, subduction zone destruction and rocks subjected to extreme pressures, forcing metamorphic realignment into new strata. That they have all been jumbled together, eventually to form the island of Anglesey is in itself intriguing. Untangling the order in which these events occurred and the logical stratigraphy which followed, is something of a challenge as the visible outcrops are by no means uniform in this respect. There are various theories, each one throwing up its own set of questions and unexplained phenomena, making the island an interesting location for ongoing geological research. The discovery of various fossils has done little to clarify the situation and recent radiometric dating, while confirming strata of up to 590 million years of age, has similarly left many questions unanswered. The precise historical relationship between Anglesey and

Fig. 3 In stark contrast to the slaty rocks of the Moel mountains

what we now know as England and Wales (Avalonia, itself formed from an arc of volcanic islands) is not absolutely clear, but it is possible that Anglesey approached Avalonia during the latter part of the Precambrian period, with its associated oceanic crust being subsumed into a subduction zone at the Avalonia boundary. Subsequently, much of this melange of crushed rocks and debris would work its way slowly back towards the surface, having been subjected to intense pressures and metamorphosed into some interesting mineral assemblages while, on the other side of the land mass, sediments were being deposited as a result of turbidity currents, creating strata of sandstones and siltstones with the occasional igneous intrusion. Such a situation might account for the variety of interesting formations to be found on the island today.

During the Ordovician period, it is likely that much of Anglesey was covered by rising seas, eroding some of the older rocks and depositing further layers of sandstones and shales. There is little visible evidence of Silurian strata, the most notable being outcrops of shales upon Parys Mountain which is of course known more for its interesting mineral deposits. Indeed, in the late eighteenth century, Parys Mountain was the largest source of copper in Europe. More recent exploration has revealed further mineral seams of copper, lead, zinc and even silver and gold. It is thought that there remain several million tonnes of ore laying deep within this area, adding further to the interesting geological picture, as details of the local stratification unfold. The return of the sea in Carboniferous times resulted in additional sedimentary layers of fossil rich limestone and mudstone being deposited, while more recent volcanic activity in the Cenozoic saw the intrusion of many mineral rich dykes, further enhancing the diversity of rock formations to be found on Anglesey.

It is interesting that, within the relatively small area of North Wales and Anglesey, there exists such a rich diversity of geological

Fig. 4 Some of the boulders deposited in the valley at Beddgelert are quite large

formations, reflecting an equally interesting geological history. Virtually every square kilometre has a fascinating tale to tell of one sort or another. No wonder it is an area which has attracted geologists over the years, both from Britain and indeed all over the world. Furthermore, from a human civilisation perspective, this interesting land has yielded rich seams of minerals and natural materials which have become interwoven with the cultural history of North Wales. The area can appear rugged and unforgiving, sometimes scarred by the activities of mankind. It can also appear extraordinarily beautiful in places, particularly in the spring and early summertime when a palette of the most beautiful colours and hues can suddenly appear within even the most rugged vista. Underpinning it all are rocks from ancient times when the world was a very different place indeed. To walk into North Wales is to walk into history, in a very real sense.

Fig. 5 Approaching the older rocks of Snowdon herself from the miners track

Fig. 6 Hard slaty rocks leading down to Lyn Ogwen

Fig. 7 A frosty Capel Curig with Snowdon rising in the background

The Peak District

The Peak District

Period - Age (MYA)	*Key*
Quaternary 0-65	
Cretaceous 65-142	
Jurassic 142-205	
Triassic 205-248	
Permian 248-290	
Carboniferous 290-354	
Devonian 354-417	
Silurian 417-443	
Ordovician 443-495	

Neoproterozoic 545-1000	
Cambrian 495-545	
Metamorphic Rocks	
Palaeozoic 500-1000	
Proterozoic 1500-3000	
Igneous Rocks	
Intrusive	
Volcanic	

The Peak District

Summary of Geology

The Peak District sits mostly upon Carboniferous limestone laid down over 300 million years ago, as is readily visible throughout the White Peak area in the south of the region. Later volcanic activity around 180 million years ago, forced rich mineral seams into weaknesses in the limestone, creating the wealth of minerals for which the area has been known throughout history. As a consequence of this activity, in combination with natural weathering and hydrological action, the area is also characterised by a number of interesting caverns and underground landscapes. The more northerly Dark Peak area is characterised by gritstone and shale overlaying the limestone. The rugged gritstone provides the basis for moorland landscapes, often retaining a great deal of moisture with peaty bogs which are to be found in many areas. Beneath all of this lies a pre-Carboniferous base which is not actually exposed anywhere within the area.

Overview

The Peak District lies in the north midlands of England, almost adjacent to north Wales and its visible form consists predominantly of Carboniferous rocks which, at around 320 million years old, are somewhat younger than the Silurian and Ordovician strata to be found in the west (400–500 million years old). To the east of the Peak District lay younger rocks from the Triassic and Jurassic periods, in a south-west to north-east diagonal seam which reaches down to the south coast around Dorset. However, deep beneath the Peak District Carboniferous shales and limestones, layers of much older Ordovician rocks have been found. These are not exposed at the surface, but have been discovered via bore holes, such as those found at Eyam, Woo Dale and Caldon Low, where rocks of Ordovician and probable Devonian age have been found. The Dark Peak area to the north of the Peak District, features much gritstone and siltstone, in contrast to the limestone of the White Peak area to the south. The limestone beds themselves are not completely uniform, as there is also much evidence of volcanic activity, with basalt outcrops visible in Millers Dale, Tideswell Dale and elsewhere. In addition, various mineral deposits have been discovered in the area, including lead which was extensively mined during the Roman occupation. Indeed, it wasn't only the Romans who mined here. Concentrations of lead within the soil often point towards mining activities in the Iron and Bronze ages as well as the later blossoming of mining during the eighteenth and nineteenth centuries. In fact, there are several disused lead mines, an interesting example being the Mandale mine in Lathkill Dale, an area mined for lead spasmodically over hundreds of years and still bearing the tell-tale scars of this activity. Consequently, the Peak District contains an interesting mix of geology and related history within its borders.

Much of the limestone would have had its origins in something akin to a tropical lagoon, around 40 km long, where sediments consisting of the remains of millions of tiny sea creatures were systematically laid down and compacted, together with fossilised crinoids and other matter. A familiar process, yet the scale of this deposition is impressive with layers of up to 600 m thick in some places. There are also intrusions of basaltic lava which are themselves up to 10 or 15 m thick, suggesting some equally impressive volcanic activity. While this environment was being formed in shallow seas, there was a great deal of geological activity in what we would now call the Scottish Highlands, with mountains being formed by colliding tectonic plates, giving

J. Ashbourn, *Geological Landscapes of Britain*,
DOI 10.1007/978-90-481-8861-1_9, © Springer Science+Business Media B.V. 2011

rise to rivers which would eventually wind their way down to these shallow waters of what we would now call the Peak District. In so doing, they would bring deposits of silt, sand and grit and overlay them upon the limestone beds, just as a river delta builds up deposits where it meets and flows into the sea. These sometimes alternating layers were subsequently compressed to form the gritstones, siltstones and sandstones of what we now know as the Dark Peak area to the north of the Peak District. The gritstones are quite course in places, with traces of small pebbles and fairly large grains of quartz or feldspar which serve to confirm their origins. Elsewhere they are worn into a visibly finer countenance, albeit from these same fundamental components.

On the eastern side of Dark Peak, features are often quite pronounced with the gritstones eroded into distinctive tors, modelled by long periods of wind, rain and frost weathering to create the landscape we see today. While this gritstone is quite hard and relatively resilient, it is often found in alternating layers of shale, which tends to be much looser in its constitution. Consequently, in some areas the entire landscape can become unstable, especially if water seeps between the layers, causing land slips. A dramatic example of this may be found slightly further west at Mam Tor near Castleton, an area where there are also numerous caverns to be found. Indeed, the area immediately adjacent to Mam Tor, while quite picturesque with its gently rolling small hills is in itself quite volatile with various depressions and channels eroded by water. Approaching the base of Mam Tor, sometimes called the shivering mountain, the rocks are unstable and loose with stones and small boulders littering the surrounding terrain. The exposed stratification is particularly interesting, with a wide granularity of individual layer depth and a particularly rugged surface.

Fig. 1 Massive limestone layers bursting from the hills at Ashford

Many of the fallen rocks remain quite fragile and are easily broken, demonstrating precisely why this area is so unstable. Indeed, Mam Tor has effectively 'eaten' the main roadway which used to run around its base, with fragments of the broken road still visible in places. A climb to the top of Mam Tor from the exposed side can be rather hazardous due to the relative instability of the surface rock which, in places, is easily dislodged by even gentle pressure. Elsewhere, it seems visibly more robust, although such an impression may be illusory given the history of this rather grand peak.

Throughout the Peak District, wherever sedimentary layers are deposited above the limestone, there are some interesting landscapes to be found, often distinguished by an equally interesting range of vegetation and associated colours of heathers, mosses and grasses. In places, the vegetation itself exhibits a sort of banding effect, no doubt influenced by the underlying geology. The gritstone found on the caps of the peaks tends to be quite dark in colour and typically gives the appearance of being dense and rugged, with many large boulders strewn over the landscape. However, some of the smaller stones are quite easily broken, revealing a much lighter interior, sometimes that of a golden, coarse grained sandstone. Such areas are typically accompanied by rugged hummocky grasses, sometimes hiding channels of water as they roll off the high ground to form streams lower down. In fact, this landscape tends to hold a great deal of moisture, creating bog like conditions in many places, interspersed with rocky outcrops which can render traversal of the area on foot quite time consuming.

The more southerly limestone areas are equally interesting, with many forms carved out of the limestone by flowing waters and erosion, sometimes forming underground caverns or deep chasms cut through the rock.

Fig. 2 Seams of flint are often found within the limestone

The most fundamental forms were created by the huge forces of glaciation during the Pleistocene cold periods or ice ages, where valleys were deepened and high summits levelled by the advancing and retreating ice. The ice also brought a good deal of debris down from the north, evidence of which may still be found in the form of erratic boulders and rocks strewn across the landscape. Towards the end of the ice age, glacial meltwaters played their part by further eroding channels and exploiting any crevasses in the limestone, creating a variety of features from small caverns to deep gorges, such as that found at the beautiful Winnats Pass, where the limestone twists and turns its way down the hill. Ancient cracks and fissures within the limestone allowed for the subsequent injection of mineral seams, as molten magma was forced up from below, creating rich veins of interesting minerals including galena (lead), calcite, barites, fluorspar, and occasionally copper. The exposure of these seams has been exploited ever since. Slightly further north, there were also occasional beds of coal created when vegetable deposits were interspersed with layers of shale, although this coal was not of a prime quality. And, of course, the limestones and shales themselves have been extensively quarried in recent times, leaving unnatural scars upon the landscape. Consequently, the contemporary landscape of the Peak District is one that has been subtly influenced by mankind, including the clearing of what would have been much natural woodland, as well as the natural forces described above.

With regard to human occupation, various archaeological remains have been carbon dated which align clearly with the Mesolithic period. This is quite interesting as we tend to consider the Mesolithic tribes as wanderers. Perhaps they wandered between favourite sites within the Peak District, maybe at different times of the year, in order to

Fig. 3 The darker, more dense stone to be found at Buxton

benefit from seasonal changes across the landscape. Curiously, there is rather less evidence discovered to date of Neolithic settlers, although we would assume that it is these Neolithic communities who first cleared the woodlands to create grazing for animals and perhaps rudimentary crop growing. Later on of course, the Romans had a significant presence around Derby and the Peak District. All of these communities would have understood and benefited from the ability of the high peak moorlands to retain water, feeding the many streams and river systems to provide a ready supply of fresh drinking water. The Romans would have also appreciated the abundance of limestone for building purposes and quickly understood the significance of the rich mineral seams to be found throughout. Many lead ingots have been found which carry Roman inscriptions, including references to the Emperor Hadrian and several appertaining to 'Lutudarum' which may well have been a small but significant town. It is perhaps not surprising that the Peak District has this long standing affinity with human activities. In spite of this fact, much of the contemporary landscape has a natural feel about it and supports a variety of interesting flora and, to a lesser extent, fauna.

While the limestone is in abundance almost everywhere, there are none the less subtle variations to be found from area to area, both in the underlying rock and the more general topography. For example, the limestone around Castleton and further north is generally of the expected medium grey colour and of fairly dense construction. Just a little further south at Deep Dale, the limestone is much lighter in colour and tends to be looser in construction and easily weathered. It also often supports a variety of lichen which lend it some additional and varied colour. Interestingly, the lichen, while present across the

Fig. 4 The paler rocks at Chee Dale adopt a variety of subtle hues

Peak District, seem to vary almost randomly in their intensity and variety. The Deep Dale area itself is quite varied, with a combination of smooth, sparsely vegetated hills and those supporting a much thicker woodland. Around Buxton, dark limestone of a particularly rugged structure is often found, although there are local variations, sometimes displaying a smoother profile with a bluish tinge and sometimes with a finer granularity of stratification. Indeed, the limestone breaks through in many areas, often with subtle differences in texture, colour and overall stratification.

Further south at Lathkill Dale, we come across a different environment with a fairly densely wooded area following the path of the shallow river as it winds its way gently through the dale. The sides of the dale are quite steep with fairly rugged limestone exposed in various places. This limestone, while superficially similar in appearance to that found throughout the area, seems to be quite dense and hard, a factor echoed by the experience of the later lead miners who found it very difficult to cut through. The steep sides seem to present no problems to the trees however, whose roots cling tenaciously to the thin soil and underlying limestone, creating a fairly dense woodland. The shallow river tumbles lazily at the head of the dale and seems to quickly grow tired as it slowly snakes its way along, diverted by human hand in several places. There are grasses along the banks of the river and even growing within the river itself. In places, it seems like the sparkling water is simply flowing over the ground. The river bed consists mostly of broken limestone pieces, still fairly angular as the water hasn't yet been able to have a rounding effect. There are one or two rogue pebbles to be found which are not of the same limestone and which have presumably been washed down from .

Throughout the Peak District, one can find much evidence of quarrying. There are several disused or dormant quarries which provide an interesting picture of the various rock strata at

Fig. 5 The rugged looking gritstones at Crowden are easily weathered and broken

the exposed face. Sometimes, there are obvious transitions with seams of sandstone and shales alternating with the limestone, sometimes almost vertical seams of slightly different texture or colouring among the limestone. Occasionally, there are even exposed crystals to be found. There are active quarries as well, providing much needed aggregates for industrial use and road building as well as agricultural lime and high purity calcium carbonate products for a variety of markets. It is interesting to consider that the Peak District has been providing raw materials and minerals for hundreds, indeed, thousands of years and is still doing so today. Contemporary technology allows such materials to be gathered and processed much more easily than in the past of course. In various areas, such as Lathkill Dale for example, there are disused lead mines where miners would spend many years slowly etching their way through the limestone in order to reach mineral seams, sometimes successfully, sometimes less so. It would have been a hard and often dangerous occupation, especially when the weather turned, as it is prone to do quite suddenly in the Peak District. The many caves and caverns provide a more pleasurable experience, with wondrous stalactites, crystal formations and unexpected colours to be seen in these underground wonderlands.

The Peak District and its geology may not, on first acquaintance, seem quite as dramatic as the Lake District or even the Yorkshire Dales, both of which paint their pictures upon a larger, more spectacular canvas. However, beneath this gentle exterior lies a wealth of interesting detail and variation which imbues the area with a distinct character of its own. This is true from both a geological and historical perspective. It may also be considered in some respects as the heart of England, straddling the southern Pennines and providing a transition to the later Triassic, Jurassic and Cretaceous landscapes in the south east of the country.

Fig. 6 Characteristic weathered limestone at Deep Dale

Fig. 7 Typical Peak District landscape at Ladybower

Fig. 8 Mam Tor, the shivering mountain

The Lake District

The Lake District

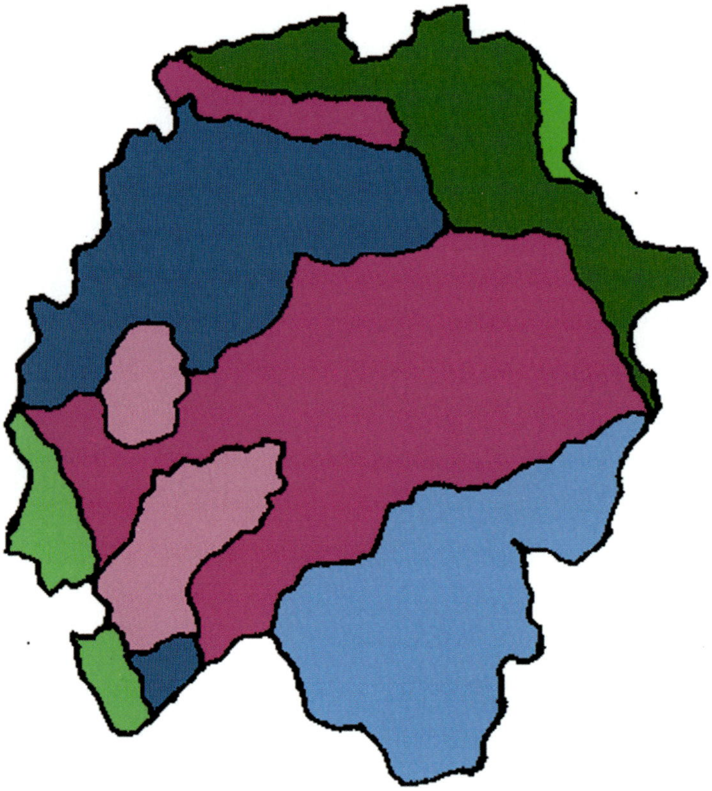

Period - Age (MYA)	*Key*
Quaternary 0-65	
Cretaceous 65-142	
Jurassic 142-205	
Triassic 205-248	
Permian 248-290	
Carboniferous 290-354	
Devonian 354-417	
Silurian 417-443	
Ordovician 443-495	

Neoproterozoic 545-1000	
Cambrian 495-545	

Metamorphic Rocks	
Palaeozoic 500-1000	
Proterozoic 1500-3000	

Igneous Rocks	
Intrusive	
Volcanic	

The Lake District

Summary

The Lake District features some of the most beautiful scenery in Britain and, being relatively compact, represents a distinct environment among the mainstream. Within this environment there exists a variety of geology and geological features, born of an interesting history which may yet be fully understood. In simple terms, one might posit that its origins lay in volcanic action and associated outcrops occurring up to 500 million years ago, juxtapositioned among slates and mudstones of a similar age and moulded by subsequent glacial action in order to form the gentle valleys and sometimes quite rugged peaks which we find today. The bulldozing glaciers mostly affected the softer rocks, scouring out the landscape and aiding the establishment of lakes and tarns in their trail within this notably wet environment. The more rugged igneous rocks tended to prevail to form the fells and peaks adjoining these valleys, ranging in height from the 2,700 ft of the High Street to the 3,200 ft of Scafell Pike. Consequently, the geological picture today is interesting and varied as one traverses this small but beautiful area of Britain.

Overview

During the early Ordovician period Britain was not as it is now, but situated in the southern hemisphere. It wasn't even a coherent whole. Northern Britain was part of the continent which became North America and southern Britain part of a different land mass situated further south and partly covered by a shallow sea. The two were separated by the Iapetus ocean, a large body of water which was to gradually shrink due to the effects of subduction, a process which gave rise to much volcanic activity around the subduction zone. By the Devonian period, around 400 million years ago, the Iapetus ocean had completely closed. Such historic activities are among those which shaped the topography of what we now know as the Lake District.

The oldest rocks tend to appear in the north and north-west of the area and belong to the Skiddaw group. These are predominantly sandstones, siltstones and mudstones originally laid down in the shallow waters of the continental slopes of the Iapetus ocean. In places, these sedimentary layers have been metamorphosed into slates, appearing as a fairly uniform grey coloured rock with, curiously, little evidence of fossils. Remarkably, these layers extend to more than 3,000 m in thickness, representing a considerable mass within this relatively contained area.

Environmental conditions evidently changed following these events, no doubt as a result of the Iapetus ocean shrinking as the associated tectonic plates moved closer together during the Ordovician period, generating a subduction zone into which the Iapetus crust was being subsumed faster than the generation of new mid-ocean crust could replenish the loss. Consequently, there was considerable volcanic activity around the edges of this subduction zone, spewing out large volumes of volcanic material. Some of this material, as expressed within the Eycott Volcanic Group, evidently was erupted under water as it is mixed in places with the Skiddaw sediments. The bulk however was erupted above water and forms the impressive Borrowdale Volcanic Group of rocks, extending from the beautiful Wasdale in the west, through Scafell and Helvellyn and across to High Street and Haweswater in the east. The extent of these volcanic eruptions is evident in that over 6,000 m of material, covering an area of 800 sq. km was erupted during a period of around 10 million years during the mid Ordovician. This represents a bout of fairly intense volcanic

J. Ashbourn, *Geological Landscapes of Britain*,
DOI 10.1007/978-90-481-8861-1_10,

activity. Furthermore, it seems that while the first stage of this activity may have been of a fairly general nature, resulting in predictable lava flows across the landscape, there was a second, more intense phase, characterised by much explosive activity with huge volumes of material erupted very quickly. Thus, we have areas of andesitic, often localised lava flows mixed with areas composed predominantly of air-fall tufts where molten andesite, dacite and rhyolite had been thrown high in the air by these enormous explosive forces before falling back to the surface. If we could have walked through the Lake District immediately after these events, we would have been presented with a quite different picture of a dramatic volcanic landscape compared with today's lush valleys and peaks.

Following these volcanic outpourings, there would have been a certain amount of natural erosion, possibly with some of the volcanic peaks collapsing. Further sediments were then lain down atop of these volcanic rocks during the late Ordovician and early Silurian periods. These varied sedimentary layers are known as the Windermere Supergroup and may be found across the Lake District, particularly in the south. There is also evidence to suggest that immediately prior to the last of these sedimentary layers there was a good deal of uplifting and folding among the earlier rocks as a result of continuing tectonic activity. There are also several igneous intrusions, ranging in size from small dykes and plugs to major plutons such as that of the Eskdale granite. Many areas of granite are now exposed due to continued erosion over the millennia, including at Ennerdale, Eskdale, Mosedale and of course at Shap where the distinctive granite has been systematically quarried for ornamental purposes. Later sediments laid down in Devonian times are rarely distinct and are mostly seen in conglomerates such as those at Mell Fell and Shap Fell, some of which contain crystals of

Fig. 1 Looking out over the valley at Boredale

orthoclase feldspar, suggesting that the granite was exposed at this time.

By the start of the Carboniferous period, continued erosion had enabled shallow seas to encroach upon the land, resulting in fairly widespread limestone deposition, followed by a series of sandstones and siltstones. The Coniston Limestone Group is indeed not formed of a single pure limestone, but rather a succession of limestones, sandstones and mudstones. This tends to imbue them with an interesting pattern of weathering as certain strata, for example those with a high calcium carbonate content, are less resistant than others. These sedimentary rocks may be found mostly around the periphery of the Lake District, having been eroded from the main core. Later sediments may well have been laid down in the Permian and Triassic periods although there is little remaining evidence of them.

And then the ice came. During the Pleistocene period Britain was subject to a series of glaciations interspersed with warmer periods. These warmer periods sometimes lasted for up to 100,000 years, but it was the glaciations which changed forever the topology of much of the country and, especially that of the Lake District. The last of the major glacial phases took place between 25,000 and 15,000 years ago with a slight reprise around 13,000 years ago in the north of the country. During these glaciations, the ice reached right down to Kent before slowly retreating again during the interglacial warm periods. As the ice retreated it scoured its way through the landscape, scraping against valley walls and creating the characteristic U-shaped valleys with which we are familiar today. In mountainous areas the rocks often show signs of glacial erosion, being both smoothed by the immense pressure of the ice and scarred by debris being dragged across the surface of the rocks. The melt-water from the ice running off to create streams and rivers, further eroding the exposed valley floors.

Fig. 2 Rydal fell quietly watching over Grasmere

The overall effect, especially in relation to the last of these glaciations, for the Lake District would be to widen and smooth the valleys while simultaneously eroding the peaks into a much smoother, more rounded profile. This has created the gentle and beautiful contours of the Lake District which provide much of the appeal of the area. This erosion has also exposed a variety of geological sequences, providing an insight into the history of the Lake District and its formation.

The lakes themselves sit in silent splendour, each with its own special characteristics formed partly by the adjacent geology. They range in size between Rydal Water with its relatively compact 31 ha to the magnificent Windermere at 1,459 ha. Ullswater, with 884 ha, sits amid a variety of interesting features and fells while Wast Water, with 283 ha sits broodingly as the deepest of the lakes at 74 m deep. From its eastern shores Scafell Pike and Great Gable rise from the mist like great hump-backed whales while the south eastern side, weathering and glacial erosion have created the relatively unstable screes which can lend Wast Water a rather foreboding air under certain weather conditions. The scenery amid the lakes and fells is truly inspiring, as has been witnessed by William Wordsworth and many others. Wordsworth lived in the vicinity of Grasmere and considered it "the loveliest spot that man hath ever found". Many of the lakes have interesting stories to tell. Haweswater, once a natural lake, was dammed and flooded in 1929 after an act of parliament granted The Manchester Corporation permission to transform the lake into a reservoir, destroying the farming villages of Measand and Mardale Green in the process and having to relocate both residents and bodies from the local churchyard. Remnants of these villages may still be observed at low water levels. Coniston of course was the scene of several world water speed record attempts, finally

Fig. 3 Wasdale Head from Wast Water

claiming the life of Donald Campbell on a bleak January in 1967 as he lost control of his Bluebird craft after exceeding 320 mph on the lake. Other lakes have their own stories, as do the surrounding fells where the Romans had several outposts and engaged in mineral extraction and various mining activities along with their more general administration. Indeed, humans have occupied these fells and valleys since Neolithic times and beyond. They have left their mark not only in archaeological monuments but, from the Bronze age onwards, in fundamentally changing the landscape by deforestation in order to provide land for grazing animals and the growing of crops. However, the fundamental character of the Lake District is more a product of those beautiful fells, valleys and of course the lakes themselves, as created by millions of years of geological activity. It is that deepest character which speaks to you most profoundly as you traverse the area. There is a sense that they have been there for a long time and that we are simply passing through.

Within this compact area of Britain which we like to call the Lake District, barely 34 mi across, there exists a wealth of both natural beauty and interesting history. It is also a most fascinating area from a geological perspective with abundant evidence of various landscape shaping processes and an attendant variety of igneous, metamorphic and sedimentary rocks. The relative impermeability of the volcanic rocks has enabled the formation of the lakes within the glacial cut valleys, creating this unique contained environment. The area represents a significant chapter in the historic geology of Britain as well as an important contemporary area within which to understand geological processes.

Fig. 4 Jumbled rocks and boulders around the shores of Thirlemere

Fig. 5 A typical Lakeland landscape at Rydal Water

Fig. 6 Looking across towards Helvellyn from High Tove

Fig. 7 Water cascading over the limestone at Aira Force

The Yorkshire Dales

The Yorkshire Dales

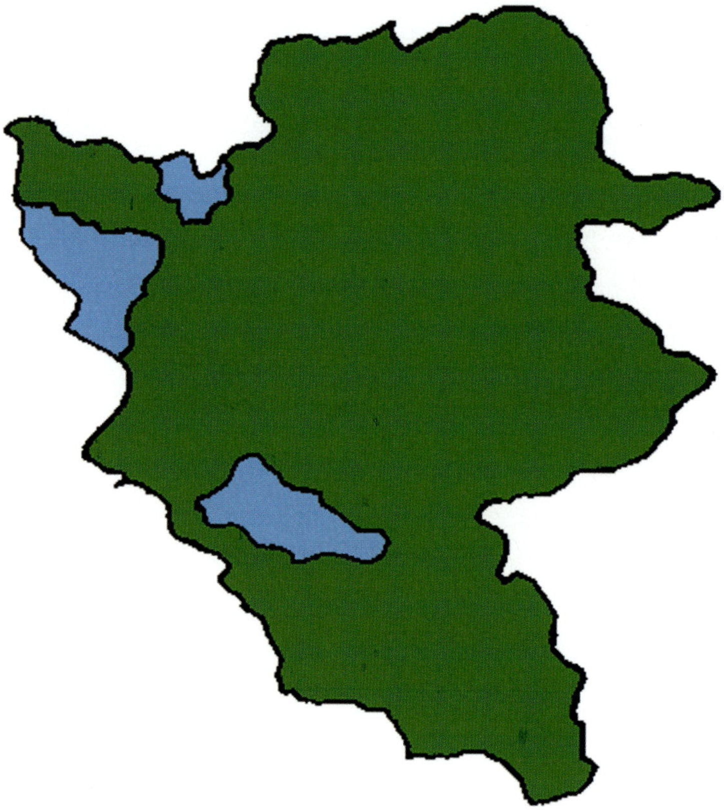

Period - Age (MYA)	*Key*
Quaternary 0-65	
Cretaceous 65-142	
Jurassic 142-205	
Triassic 205-248	
Permian 248-290	
Carboniferous 290-354	
Devonian 354-417	
Silurian 417-443	
Ordovician 443-495	

Neoproterozoic 545-1000	
Cambrian 495-545	

Metamorphic Rocks	
Palaeozoic 500-1000	
Proterozoic 1500-3000	

Igneous Rocks	
Intrusive	
Volcanic	

The Yorkshire Dales

Summary of Geology

The Yorkshire Dales are founded primarily on Carboniferous limestone laid down around 300 million years ago. The Great Scar limestone tends to be dominant, especially around the Ingleton and Settle area, where it attains a thickness of around 200 m, displaying some beautiful limestone pavements at the surface. Above, and sometimes alternating with this limestone on the higher ground, sits the Yoredale Series of gritstones, sandstones and shales which, due to their varying granularity and weathering properties, often create a stepped landscape as well as supporting moorland environments. Fault lines create some distinctive features, while the whole of the Yorkshire Dales landscape is very much influenced by the last ice age, with many glacial features in evidence, from the broad 'U' shaped valleys to the presence of erratics and redirected streams and rivers.

Overview

The Yorkshire Dales are considered as predominantly limestone country, yet there are interesting variations as one moves from one area to another, including the presence of shales, gritstones, sandstones and even granite. Actually, the landscape we see today owes much to the last ice age where glacial action scoured the valleys into their familiar gentle U shaped profiles, leaving the higher ground either side. Beneath this beautiful vista, and occasionally breaking through it, lies a massive swathe of Carboniferous limestone, aligned in a south west/north east diagonal across the country. The combination of rivers, streams and precipitation have variously interacted with this limestone, causing a good deal of natural weathering, resulting in the scars, limestone pavements and many underground chasms and caves which characterise this part of the country.

Towards the north, Swaledale and Arkengarthdale feature a distinctly banded geology with alternating layers of limestone, sandstone and shale, some of them no more than 20 m thick. These layers are known as the Yoredale series, which tend to be darker than the limestone found further south and which support a variety of natural flora, lending a distinctive appearance to this area. On the higher ground around Swaledale the landscape approaches a moor like appearance in places, with rugged grasses and mosses in abundance. However, the uppermost layer of limestone breaks through almost everywhere, sometimes in an a seemingly haphazard fashion, no doubt enhanced by weathering, as can be seen at places such as Napa Scar, where the rocks have become unstable and liable to crumble and slide down the valley slopes. The limestone at Napa Scar is fossil rich and almost every piece you look at contains evidence of an abundance of past marine life, suggesting origination in a warm shallow sea. It is also quite fragile in places and easily broken, as evidenced by the avalanche of small rocks flowing down from the main scar area. The Yoredale limestones tend to be slightly darker in appearance to the limestone found further south, often referred to as the Great Scar limestone.

In some places, the Yoredale limestone is covered by a series of quite coarse shales and sandstones known as the Millstone Grit. This strata of gritstone tends to cover the high ground either side of the dales, supporting the peaty moorland plateaus of cotton grass and heather, often with depressions and small mounds creating a varied and attractive landscape. These higher plateaus, such as that atop Swaledale, are quite exposed and subject to the full force of the weather, which can be fairly extreme at times. It is also on these plateaus

J. Ashbourn, *Geological Landscapes of Britain*,
DOI 10.1007/978-90-481-8861-1_11,

that the river Swale has its origins, around the area known as High Seat. From here, it quickly gathers momentum and winds its way, via a series of sometimes spectacular waterfalls, through Keld and on to Richmond Bridge and beyond. In fact, the course of the river was modified by the last ice age when glacial debris blocked parts of the original flow, while glacial meltwater served to cut new gorges and divert the flow to that of its present course. Glacial retreat moraines are also to be found in various locations, providing more evidence of significant glacial activity.

In places, particularly around the western side of the dales, one is often surprised at the rather extreme angle of stratification to be found within the limestone beds. This is a consequence of a major geological contrast between the Howgill Fells in the Lake District and the Yorkshire Dales, marked by a 32 km fault line known as the Dent Fault. It was formed when the Silurian rock of the Howgill Fells rose up spectacularly to create the mountainous landscape we see today, pushing against the limestone scars of the Yorkshire Dales until the edges of the limestone were forced up into a near vertical position adjacent to the fault line. Much of the Great Scar limestone has been affected around Dentdale, Garsdale and Barbondale with reverberations elsewhere, although the relatively flat valley bottom near Gawthrop is thought to have once been a glacial lake formed during the last ice age. Variously sitting above the Great Scar limestone are of course the Yoredale series of alternating layers of limestones, sandstones and shales.

Wensleydale is a particularly interesting area, comprising of a long, wide dale rising gently to the surrounding hills, but often with some hard edged steps breaking through the otherwise smooth landscape. These are a result of harder limestone layers among the more easily eroded softer sandstones and shales of the Yoredale series. At the heart of Wensleydale runs the

Fig. 1 The edge of a limestone pavement reaching out like fingers at Ingleton

river Ure, which starts life high on the borders of Cumbria and winds its way along the dale, sometimes flowing gently within a peaceful vista, sometimes babbling its way over coarser ground and sometimes taking the form of waterfalls as it traverses the steep steps of limestone. Throughout its course, the Ure drops from a level 305 m above sea level at the dale head, to around 90 m above sea level at Kilgram Bridge, down at the lower end. At Hardraw Force, the waters of Fossdale Gill run down from the hills where they drop down 27 m over the hard limestone edges to join this spectacular composition. Such waterfalls are often the result of re-routed flows after glacial retreat left debris, known as glacial tills, in its wake. This diversity of form is matched by a diversity of underlying geology. Deep beneath Wensleydale, around 500 m below the surface, lies a pinkish granite around 400 million years of age, on top of which lay the 350 million years old Great Scar limestones, followed by the Yoredale series of limestones, sandstones and shales. The Yoredale layers vary in thickness at different locations, exposing different strata to create this interesting landscape. Around Askrigg for example, the Yoredales are themselves around 275 m deep, while considerably less so elsewhere. There are also limestone pavements to be found, such as around Aysgarth, where the ground rose up, causing the river to tumble its way down the steps of alternating limestone and shale, as if walking down a staircase. Contrasting layers make up the river bed and the sides of the channel, creating an interesting and, in places, quite beautiful landscape. Just south of Wensleydale lies Semer Water, an unexpected, quite large glacial lake nestling in a hollow plateau where the wind can sometimes drop down and create abstract designs upon its surface. It was once much larger and the overflow from the lake formed the gorge and associated waterfalls of the river Bain, the shortest river in England, which flows

Fig. 2 Chasms open up in the limestone to underground caves and rivers

above Bainbridge. Move northward from Askrigg and you climb up onto the moorland plateau, with its mosaic of scattered limestone boulders, which marks the transition into Swaledale.

Moving further south to the Craven Dales and the Three Peaks of Whernside, Pen-y-gent and Ingleborough, the scenery is quite spectacular with the contrasting peaks and broad dales formed of the Great Scar limestone and subsequent Yoredale series, often resulting in a distinctly stepped appearance. Fault lines, such as the South Craven Fault near Settle, sometimes provide a contrast between the limestone and gritstone scenery but, generally, the limestone dominates the perceived picture. There are many beautiful limestone pavements in the area with particularly interesting examples around Ingleborough. Just east of the Ribblehead Viaduct, high on the hill, the limestone pavement is especially rugged and impressive, with deep crevasses carving an intricate pattern among the limestone blocks. There are also a number of limestone boulders strewn around in the vicinity and a great many shake holes. At the end of the pavement, one can hear rushing water, even though there is no visible stream. The sound is coming from a dark and gloomy looking adjacent hole which reaches down to an underground stream. Such underground caverns and streams are common in the area, caused by water wearing away the softer layers of rock, often creating some spectacular and beautiful underground scenes. They can also be quite dangerous, especially when entry points are exposed to the surface, but partially hidden among the long grass. In some areas, such as at Gaping Gill, the entrance shafts are obvious and much larger. Indeed, Fell Beck plunges 103 m into Gaping Gill, effectively forming one of Britain's largest waterfalls. The area of the Craven Dales is marked by picturesque rivers, magnificent

Fig. 3 Limestone riverbed at Langstrothdale

peaks, gently undulating countryside with beautiful shades of green, brown and yellow and, of course, limestone. A great deal of limestone in fact, although the gritstones and shales of the Yoredale series are also present on the higher ground.

Malhamdale is a popular area for those interested in geology, with some distinct features among the Great Scar limestone which dominates the area. Many of these features result from glacial activity during the last ice age. Malham Tarn, is a glacial lake sitting in a basin scoured out of the landscape by ice. The area is surrounded by an interesting landscape, including bogs, gently rolling hills, limestone scars and limestone pavements. Faults such as the North and Mid Craven Faults have contributed to the landscape, shaping the limestone wall of Malham Cove and the Gordale Scar among other features. Erosion of the softer shale to the south of the Mid Craven Fault line has created a distinct step between this rock and the harder limestone plateau to the north of the line, accentuated no doubt by further erosion during the ice age. The Gordale Scar is an impressive gorge cut through the limestone by glacial meltwater, leaving a towering chasm almost 100 m high, reminding us of the power of these natural processes. Indeed, the limestone around the Malham area has been cut, shaped, eroded and crumpled by a succession of events, leaving behind a very distinctive landscape of stepped valley walls, limestone pavements and many limestone boulders of various sizes scattered around, some of the larger ones showing their own particular history of weathering over time. One gets the impression that this is a landscape created by some powerful unseen hand, in a bygone drama.

By comparison, the landscape around Wharfedale seems gentle and relaxed, with broad, smooth valleys crafted by glaciation and subsequent melt periods. The Great Scar

Fig. 4 Typical Dales landscape at Wensleydale

limestone is topped in places by the Yoredale series, especially in the upper Wharfedale area north of Kettlewell, where various outcrops may be found. Littondale and Langstrothdale reach out at a tangent towards the north west and feature interesting landscapes of their own. Langstrothdale in particular has a series of shallow limestone terraces over which the waters cascade to eventually join the main river Wharfe just north of Buckden. Further along, at Langstrothdale Chase, tributaries feed in from the higher ground towards Wether Fell, where the landscape changes to more of a moorland plateau before rolling down again towards Hawes in Wensleydale. But Wharfedale has all the attributes of typical limestone scenery including small gorges, caves, limestone pavements and even shake holes in places, reminding us that there is an additional underground landscape of streams and caverns, with their own little waterfalls and distinct features carved out of the limestone.

Throughout the Yorkshire Dales, there is are a variety of individual landscapes to be found as one traverses the various peaks and dales. The higher ground often featuring beautiful plateaus of moorland and distinctive stratification of the limestone, gritstone and shales, sometimes stepped in gigantic terraces upon the hillsides. The lower ground, through the valley floors can be equally varied, from smooth wide channels to more erratic landscapes, many of which owe their visible form as much to the hand of man as to natural processes. The transition between the two is often quite steep and there are many hills to climb in order to witness it. There are also many delightful little spots where one can find the unexpected. At the southern end of Coverdale for example, lie some little streams where the expected limestone boulders and pebbles are occasionally interspersed with some well rounded sandstones, no doubt systematically eroded and

Fig. 5 An avalanche of limestone at Napa Scar

washed down from the higher ground over time. Further up the dale are some unusual land formations whose contrasting colours and shapes reflect the underlying geology and the processes which have acted upon it over a very long time. There is also a human influence, as settlers have populated the dales throughout the past 10,000 years, leaving clues behind them of ancient settlement sites, later mining activities and of course the intricate patchwork of dry-stone walls which is so characteristic of what we see today. It is interesting to ponder upon what had once been the bottom of a far away marine lagoon, and yet has witnessed so much activity to become the beautiful landscape of the contemporary Yorkshire Dales. The journey continues of course, and one wonders what this landscape will look like in a million or so years time.

Fig. 6 Limestone beds at Aysgarth

Fig. 7 Well worn pebbles in the valley at Swaledale

Fig. 8 Water cutting through the limestone at Aysgarth

The Yorkshire Moors

The Yorkshire Moors

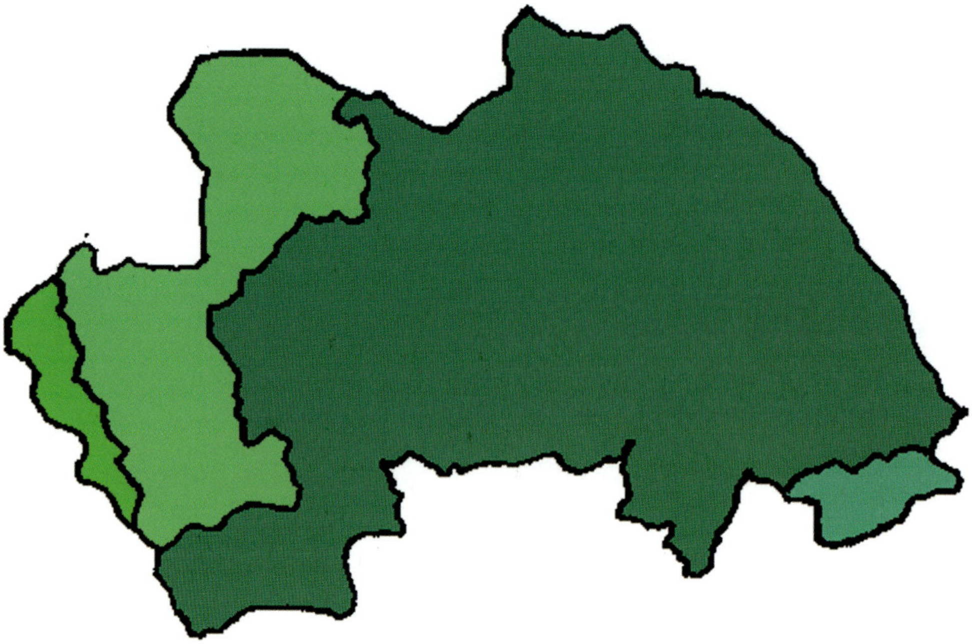

Period - Age (MYA)	*Key*
Quaternary 0-65	
Cretaceous 65-142	
Jurassic 142-205	
Triassic 205-248	
Permian 248-290	
Carboniferous 290-354	
Devonian 354-417	
Silurian 417-443	
Ordovician 443-495	

Neoproterozoic 545-1000	
Cambrian 495-545	

Metamorphic Rocks	
Palaeozoic 500-1000	
Proterozoic 1500-3000	

Igneous Rocks	
Intrusive	
Volcanic	

The Yorkshire Moors

Summary of Geology

The geology of the Yorkshire moors marks a transition from west to east as the older carboniferous rocks found in the dales to the west give rise to the younger Triassic and Jurassic rocks in the east. The moors consist almost entirely of Jurassic sedimentary limestones, sandstones, gritstones and shales, mostly from the middle Jurassic period, around 170 million years of age, although rocks from the early Jurassic period, around 200 million years of age are also exposed in places along the coastline. Glacial features are also to be found, the landscape owing much of its shape to the last ice age and subsequent meltwater activity. It is perhaps a gentler more flowing landscape than the dales to the west, but of course also has an adjacent coastline which is interesting in itself.

Overview

The Yorkshire moors are characterised by the higher upland plateaus whose distinctive and beautiful swathes of purple heather and golden grasses perhaps typify this kind of landscape for many admirers. Interestingly, there is also a good deal of woodland, with over a fifth of the area populated by a variety of trees including oak, ash, birch and rowan which, in turn, support their own habitats and associated wildlife. It is an area rich in treasures for the naturalist and all those interested to discover how nature finds a way to enable species to work collaboratively within such a wonderful environment. However, it is also an area rich in geology and associated history, providing an interesting example of how such a landscape comes into being.

The central moorland area rises up to the north and west, marked by quite sharp divisions and escarpments, betraying a history of erosion around this perimeter by the continual effects of weather and river action. To the south, the transition is more of a gentle slope as the landscape tilts downwards towards Pickering. To the east, we have a varied and often dramatic coastline, displaying a mixture of interesting geology, including the oldest rocks to be found in the area which date from the early Jurassic period, around 200 million years of age. These variously coloured shales, sandstones and limestones are often rich in fossil remains of ammonites and other marine organisms and, of course, the area is also well known for its dinosaur footprints, often being called the Dinosaur Coast. This reminds us of the marine origins of this landscape although, by around 170 million years ago, the sea level had fallen and given way to swamps interspersed with rivers, creating the sandstone layers observed above the earlier shales along the coastline and which form the core of much of the central moorland. During the late Jurassic, around 150 million years ago, sea levels rose again, depositing strata of limestone on top of earlier layers, as often observed on the higher ground. Much of the landscape, as is mostly the case in the north of the country, has subsequently been shaped by glacial action during the last ice age, when the area we now call the moors was bordered on three sides by huge ice sheets hundreds of metres thick, but it also bears the traces of human mining and quarrying activity, courtesy of the minerals and raw materials to be found throughout the area.

A distinctive feature chiselled out during the ice age is a steep escarpment which rises at Sutton Bank to a height of around 140 m and continues right along to the coast at Scarborough. Akin to an inland cliff, this notch in the landscape is topped by a hard gritstone which is also found on top of the cliffs at Scarborough. Lower down the cliffs are softer layers of clays and limestones from the late Jurassic, beneath which sit

J. Ashbourn, *Geological Landscapes of Britain*,
DOI 10.1007/978-90-481-8861-1_12, © Springer Science+Business Media B.V. 2011

the harder mid Jurassic sandstones and then the softer shales and mudstones from the early Jurassic. Consequently, more than 60 million years of history is exposed for us within the strata of these cliffs. Below Sutton Bank lies a glacial lake named Gormire, formed when the ice melted, which also serves to remind us of the dramatic activity of these times.

The Cleveland Hills, to the north and west of the moors, feature a steep incline on their northern slopes and a softer, more gentle slope to the south, giving them a distinctive and impressive appearance. Their overall shape is a result of the previously mentioned heavy erosion along the northern and western boundaries of the moors. In fact, one of these hills, Roseberry Topping, was separated from the main body by this erosion and now stands as a distinctive feature with a layer of hard mid Jurassic sandstone forming its summit. Remarkably, there was an injection of igneous rock into the area around 58 million years ago, almost certainly as a result of volcanic activity further north in Scotland. This created a seam named the Cleveland Dyke which has subsequently been exploited for road stone purposes, leaving something of a man made scar upon the landscape. Other channels have more natural causes, such as the Newtondale Gorge which was caused by meltwater from overflowing glacial lakes. Indeed, the upland moors are intersected by a series of dales created in this way as water rushed down from the northern plateaus. Today, some of them have small becks or streams running along the bottom of the dale, reminding us of this earlier drama.

Further south, the Tabular Hills are formed from relatively young late Jurassic limestones around 150 million years of age. They feature gently sloping flat tops which dip downwards towards the south, caused when the landscape was uplifted and tilted during earth movements around 30 million years ago, exposing

Fig. 1 The Cleveland Hills rise gently from the moors

older rocks in the north. In fact, much of the Yorkshire Moors is tilted in one direction or another as the northern moors tend to exhibit a tilt towards the east. The gentle slopes of the Tabular Hills are occasionally interrupted by gorges and valleys caused by glacial meltwater run-off. One peculiar feature is the Hole of Horcum, a scallop shaped valley formed by the erosive action of multiple springs, while a more significant escarpment marks the boundary of both the Tabular Hills with the central moorland and rocks of the middle and late Jurassic. In the south west, the Hambleton Hills rise up east of Thirsk to mark the beginning of the Yorkshire Moors landscape, from where the central moorland continues, with its porous sandstone enabling the free draining qualities which support the distinctive heathland vegetation.

To the east, we have the rugged, but volatile north Yorkshire coastline, formed predominantly of early and middle Jurassic sediments, whose variations lead to some distinctive erosion. To the north of Whitby, the early shales and mudstones form a pavement which may easily be discerned at low tide, while the adjoining cliffs display a variety of strata, sometimes quite fine in its granularity, at other times more rugged and coarse in texture. There are some subtle and interesting hues among the slaty mudstone and sandstone layers, creating an unexpected but beautiful vista across the surface of the cliffs. This combination of distinctive cliffs and mudstone pavements is further accentuated by a swathe of boulders and pebbles of varying dimensions and colours, creating a spectacular scene against the more uniform hues of the adjacent sea. Some of the larger boulders are quite dark in colour, others appearing almost golden in the afternoon sun and others a pale grey. The smaller pebbles reflect this diversity, with many interesting shapes and colours created over time by erosion and wave action. Erosion has

Fig. 2 Boulders dislodged by the waters running off the moors

also shaped the profile of the cliffs, with shallow scalloped coves, together with assorted smaller features distributed across the cliff face. Indeed, some of the cliff surfaces appear to be somewhat unstable, with loose chunks of rock protruding from the surface and occasionally being dislodged altogether. They also tend to be wet with groundwater seeping through from inland, while being subject to further erosion from the waves at high tides.

Along the coast at Old Nab, an interesting strata of fossiliferous shales, sandstone and ironstone (known as the Cleveland Ironstone), stretch out between here and Staithes, providing a subtle contrast of textures and hues. Closer to the town, at Sandsend, the coastal landscape temporarily adopts a dramatically different appearance, almost other-worldly, in the form of low, soft sandy cliffs which look particularly vulnerable and, indeed, have been shored up in places. However, this is not so much a natural feature, but the result of human exploitation of the alum-rich lower shales, resulting in extreme quarrying action. Where the shale and sandstone cliffs still exist, they are often marked by bore holes left by humans mining for jet, created by the fossilised branches of Jurassic monkey puzzle trees. Overall, it is an interesting coastline, exposing a broad range of sedimentary layers, some of which are quite rich in fossils. However, while appearing quite rugged, the cliffs are none the less vulnerable and subject to erosion, as can be clearly observed in many places.

Returning west to the moors, the gently sloping hills and vales support a variety of rich vegetation, leading to some dramatic and colourful vistas which change their hues with the seasons. Sometimes a blaze of purple heather and golden grasses, sometimes a more rustic blend of yellows and browns, with woodland swathes cutting across the landscape. Underlying this topography are predominantly

Fig. 3 The hills adopt some distinctive shapes at their lower extremities

limestones and sandstones although, particularly in the south, clays are also to be found, providing a colourful base where exposed. The various streams and becks can also be quite colourful as they wind their way along, often bubbling over small boulders of limestone or gritstone and sometimes bringing an assortment of pebbles along with them. Sometimes, they expose the local sedimentary strata along their banks as the waters relentlessly carve the course of the stream into the landscape. Their paths are often lined with trees, creating a habitat supportive of a variety of wildlife. The flow of such streams is variable according to prevailing weather conditions, sometimes little more than a trickle, sometimes a veritable torrent as the waters flow down from the higher moorland plateaus. At such times, the water can adopt a muddy complexion as it erodes a little of the surrounding rocks and clay. The Yorkshire Moors certainly present a most interesting landscape of varying geology and natural habitats. They additionally present a fine example of geology in action and how various strata have been deposited, manoeuvred, uplifted and finally sculptured by glacial action to form the characteristic and beautiful environment we see today.

Fig. 4 Rugged limestone exposed at the stream bed

Fig. 5 A fragmented mudstone pavement on the beach at Whitby

Fig. 6 Clear and relatively fine stratification within the mudstone cliffs

Fig. 7 An interesting combination of textures along the Yorkshire coast

Scottish Lowlands

The Scottish Lowlands

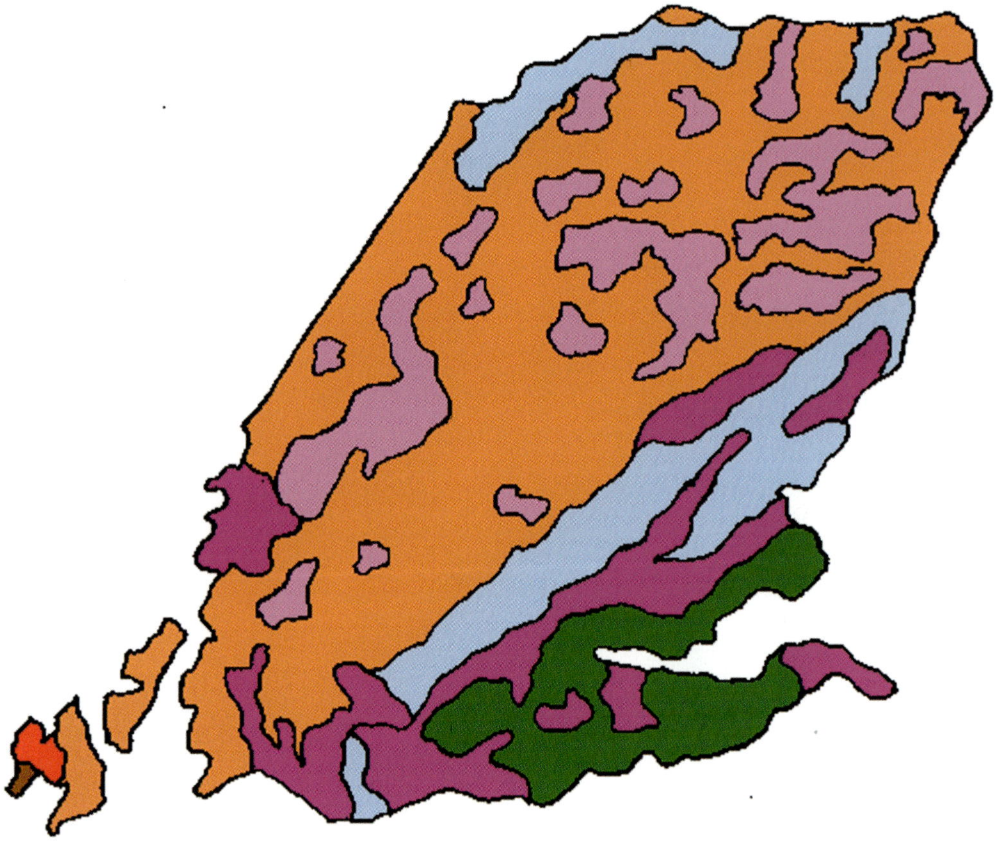

Period - Age (MYA)	*Key*
Quaternary 0-65	
Cretaceous 65-142	
Jurassic 142-205	
Triassic 205-248	
Permian 248-290	
Carboniferous 290-354	
Devonian 354-417	
Silurian 417-443	
Ordovician 443-495	

Neoproterozoic 545-1000	
Cambrian 495-545	
Metamorphic Rocks	
Palaeozoic 500-1000	
Proterozoic 1500-3000	
Igneous Rocks	
Intrusive	
Volcanic	

Scottish Lowlands

Summary of Geology

The geology of Scotland includes some of the oldest rocks to be found anywhere and certainly the oldest in Britain, with rocks from the lower Proterozoic and Archaean to be found in the extreme north. The strata generally traverses Scotland in a south-west to north-east diagonal and, in the lowlands this takes the form of a band of Devonian rock, 354–417 million years of age, stretching across from just north of Glasgow in the west to Dundee and north of Dundee in the east. Above this band, the bulk of the lowlands are characterised by underlying rocks of the lower Palaeozoic and upper Proterozoic between 500 and 1000 million years of age. These are interspersed with intrusive igneous rocks across the area and, in the west there are rocks of volcanic origin which also occur in a thin band beneath the Devonian swathe, reaching across from Glasgow to Dundee.

Overview

The Scottish lowlands are separated from the highlands by the Highland Boundary Fault which runs right across Scotland from Arran to Stonehaven, separating the rugged terrain of the highlands from the gentler rolling hills which stretch into the border counties and England. Along this dividing line lies the Highland Border Complex, a relatively thin line of rocks which have revealed some interesting fossils of trilobites, brachiopods, gastropods and other marine life which are subtly different from others found in England and Wales from around the same period. Such finds serve to confirm the view that these rocks were once part of a different continental land mass, supporting the theory that the origin of the highlands lies within Laurentia, the land mass which included parts of North America and Scandinavia, while England and Wales originated on the other side of the Iapetus Ocean within Avalonia and Baltica. The lowlands are founded predominantly upon Devonian and Carboniferous rocks from 290 to 417 million years of age.

The Isle of Arran is a southerly island of particular interest from a geological perspective as it encompasses a broad variety of rocks and features. The underlying strata are Precambrian metamorphic rocks of over 500 million years of age. These are overlain by more recent sedimentary rocks from the Lower Carboniferous period around 320–350 million years of age, suggesting some interesting historical activity which concluded in the whole being uplifted and exposed as an island. There are also Tertiary igneous rocks and granite intrusions with a series of Tertiary dykes, or dyke swarms, intruding into the later sedimentary layers. Little wonder that Arran has interested geologists since the nineteenth century, with many coming to study this most interesting and varied landscape.

Given its latitude, it is perhaps unsurprising that much of Scotland's landscape has been shaped by glacial activity, at times being completely covered by ice and, at other times, subject to the gouging effect of glaciers as they advance and retreat, cutting their way across the landscape and leaving their characteristic trails. The Cairngorm Mountains reflect such activities from the last ice age which was at its peak around 18,000 years ago, with glacial activity extending to around 10,000 years ago. The Cairngorms were originally formed by a massive magma intrusion into the crust during the Devonian period around 417–354 million years ago. The resulting granite was subsequently exposed at the surface around 50 million years ago as erosion wore away the softer sedimentary layers, revealing the rugged body of the underlying mountains. Subsequent glacial activity has further shaped the landscape,

J. Ashbourn, *Geological Landscapes of Britain*,
DOI 10.1007/978-90-481-8861-1_13,

creating the valleys and tors visible today, together with the erratics, meltwater channels, moraines and other features associated with such natural forces. The Cairngorms capture this relatively recent geological development in a picturesque snapshot which nevertheless seems very much alive, the brooding rock masses giving the impression that they are still engaged in some important journey across time and space.

Further evidence of glacial activity is provided at Carstairs Kames with an excellent example of an esker system, formed by meltwater sediment being laid down as the glaciers retreated towards the end of the last ice age. The resulting landscape is a series of gently winding ridges and mounds stretching out northwards from Carstairs itself. The appearance is almost that of a giant river bed, shaped by the flowing waters of some raging torrent and, indeed, such an intuitive perspective is not far from the truth as it was the flowing and turbulent meltwaters, running down from the higher ground which carried with them an enormous amount of sediment to be deposited as the mounds and ridges which we see today. Some of these ridges are around 25 m high which, in turn, sit upon a bed of sediment approaching 40 m deep in places, giving an idea of the scale of these past activities. This sedimentary layer tends to be graduated with coarser, boulder like deposits at the bottom, with much finer layers of gravel and sandy deposits towards the top.

There is also evidence of much volcanic activity, with Glen Coe offering an interesting example. This picturesque peaks and valley area consists predominantly of lava flows which were once associated with a larger caldera, sometimes active, sometimes not. Consequently, there are sedimentary layers among the basaltic and andesitic lavas, creating a most interesting and quite beautiful overall structure. At Glen Tilt, a different picture

Fig. 1 Signs of compression in the rocks at Arrochar

emerges where the host sedimentary layers are interjected with a number of granite intrusions, forcing their way through and among the host rock as if powered by some magnificent underlying force. Such structures evince a picture of volatile activity as the very fabric of the land was hewn and fashioned over time into the sleepy landscapes we gaze upon today. The impression is heightened at locations such as Siccar Point on the coast where sedimentary strata which are already pushed up and tilted by a great unseen force are overlain by further sediments, cutting across in a somewhat haphazard unconformity. These later layers, comprised mostly of Carboniferous sandstones, look almost as if they have been dropped onto the existing landscape and left to find their own position within the broader scheme of things.

Within this interesting geological tapestry, there are occasional veins of various ores weaving their way through the picture, as has been understood and exploited for hundreds of years. The Leadhills area, as the name suggests, has been a centre of lead mining and was for many years considered the primary ore deposit in Scotland, with much activity between the seventeenth and early twentieth centuries, although it is thought that mining in the area might go back to Roman times. Copper and zinc were also found here, as was silver and gold, among a rich concoction of mineral veins running in a north-west, south-east direction. In fact, over 70 minerals have been found among these amazing deposits, laid down in two phases, firstly during the Silurian period, over 400 million years ago and a later metalliferous phase during the Carboniferous period, around 300 million years ago, as nature completed this interesting creation.

Such interesting activities have not been limited to inland areas. Coastal landforms have also been developing in an active manner,

Fig. 2 Sedimentary layers high up in the hills

albeit in more recent time periods measured in thousands of years. The sands at Forvie, just north of Aberdeen stretch for around 24 km in an interesting complex of gently undulating golden dunes, driven by coastal processes. The Forvie peninsula and Forveran links are separated by the estuary of the Ythan river, with two other major contributing estuaries, those of the Dee and the Don completing the picture. The sands would have originally been created as glacial meltwaters brought vast quantities of sediments down to the coast at the end of the last ice age. They would have been dispersed reasonably well into the sea, although the consequent rising sea levels due to the ice melt, ensured that much of this sediment was pushed back against the coastline. A combination of the actions of waves, tides and wind has subsequently created an interesting landscape of sand dunes. However, it is a dynamic landscape which remains active as this elements continually shape and reshape the shifting sands. But not all of the coastal areas are like this. At Girvan, quite a different picture is presented, with particularly rugged sedimentary layers of between 450 and 490 million years of age. Indeed, the coast between Girvan and Ballantrae is particularly interesting as it features an ophiolite complex whereby fragments of oceanic crust have become uplifted and added to the coastal geology. A similar phenomenon occurs at the Lizard in Cornwall. The Girvan to Ballantrae complex exhibits strata from deep within the crust, possibly from a subduction zone associated with the Iapetus ocean. Shales, gabbros, cherts and pillow lavas are all in evidence and take some distinctive forms including some quite extreme folding in places. It is indeed an interesting area.

The island of Jura, just off the west coast of Scotland is characterised at first sight by the Paps of Jura, three mountains between 734 and 785 m in height which immediately take

Fig. 3 The rugged landscape of Ben Lawers

the eye of the visitor. However, it is perhaps the west coast of Jura that is of particular interest from a geological perspective as it features some very distinctive raised shore platforms, appearing almost like terraces overlooking the sea. These have been created as a result of repeated cycles of glaciation and melt during the Quaternary ice age, wherein the relative sea level was rising and falling accordingly, periodically exposing rock along the shoreline to the ravages of cold climate processes. The situation was no doubt accentuated during the warmer periods by a degree of isostatic rebound as the land was released form the burden of the weight of the ice sheets. Indeed, relative sea level in the area was at its highest around 15,000 years ago. The west coastal landscape features a distinctive layering of low, main and high rock platforms which appear almost sculpted and within which are contained various smaller terraces and ridges. There are also raised beaches of shingle ridges sitting atop these layers in places along the coast, providing a rare example of such glacial inspired features. The high and low shingle ridges reflecting the relative rise and fall of sea level, although there is a bias in altitude towards the north suggesting a more pronounced isostatic rebound in this direction. Some of the shingle ridges, notably those overlying the main rock platform, were probably developed in the postglacial period around 7,000 years ago. While the shingle beaches are unvegetated, the island overall supports a rich variety of flora and fauna and is known for its wildlife. Interestingly, there is also evidence of early human occupation, dating right back to mesolithic times.

On the other side of Scotland in the Moray Firth area, a different shingle landscape is to be found at the estuary of the river Spey. The Spey Bay area is well known for its combination of active and relic shingle ridge activity.

Fig. 4 A broad calcite seam among the twisted sediments on Ben Lawers

Towards the end of the last ice age, significant quantities of shingle were deposited in the area, partly from sediments washed down into the estuary from higher ground as a result of melt and partly from shingle being washed onshore as a result of rising sea levels as the Spey estuary was flooded. As sea levels fell and the floodwaters receded, shingle ridges were developed along the coastline as a result of fluvial processes. While the orientation of later shingle ridges has changed slightly, the processes continue over a stretch of around 8 km of coastline, interacting with natural coastal processes and creating some distinctive features, including an active shingle spit. The area is also interesting from a coastal landform perspective and features several small areas of saltmarsh and offshore deltas, supporting a variety of wildlife and, of course, the river Spey is noted for its salmon.

The Lowlands and Southern Uplands are also important from a palaeontology perspective with fossil finds throughout the area helping to corroborate the likely geological development. One of the first to take this approach was Charles Lapworth (1842–1920) who, while studying graptolite fossils, noticed distinct changes in fossil appearance within the various strata of exposed shales, enabling him to construct a plausible sequence. As a result of his pioneering work, graptolite fossils have become a common mechanism for correlating Ordovician and Silurian strata. One of the sites where this is evident is at Dob's Linn in the central Southern Uplands where the Ordovician/Silurian boundary is clearly exposed via the Moffat Shale group of rocks, a sequence around 100 m thick thought to have been originally deposited in the deeper waters of the Iapetus Ocean. Interesting fossil finds have also been made at Lesmahagow where a section of Silurian sediments, up to 440 million years of age, are surrounded by sediments from the later Carboniferous period

Fig. 5 The hard slaty mudstones at Dochart

around 300 million years ago. The older rocks consist predominantly of shales and sandstones with occasional coarser pebble conglomerates, suggesting a transition towards shallower waters. This is reflected in the fossil finds, wherein the lower strata have been found to contain marine fossils such as trilobites and brachiopods, while higher sections tend to contain evidence of river and estuary conditions.

Perhaps one of the most interesting areas for fossil finds however is Rhynie, where the layer known as the Rhynie Chert has revealed some quite remarkable early fossils including well preserved plants and arthropods. The completeness of these early fossils is partly a result of the siliceous sinter within which they are embedded, caused by hot spring activity during the Devonian period. The same processes lead to some interesting mineral seams, including gold, although not in any significant quantity. The Rhynie Chert itself varies in appearance, mostly of a blue-black colouring but with occasional lighter areas of a metallic sandy appearance. The contained fossils are beautifully preserved and include some of the earliest vascular plants as well as some of the earliest lichen found anywhere, together with various blue-green algae, rendering the Rhynie Chert both unique in appearance and uniquely valuable from an palaeontology perspective.

In conclusion, the Scottish Lowlands provides a rich variety of geology and geological structures and landforms, both within coastal areas and further inland. Some of the underlying rock formations are more than 500 million years of age, while much younger layers are also to be found in many areas. Sedimentary and igneous rock types are to be found, together with a rich fossil history in many areas. All of this is beautifully presented within a landscape which has charmed both geologists and more general visitors for many hundreds of years.

Fig. 6 Big boulders block the stream at Falloch

Fig. 7 Slaty outcrops at Loch Long

Fig. 8 Typical landscape at Loch Lubhair

Fig. 9 A world of history in the sediments at Loch Long

Scottish Highlands

The Scottish Highlands

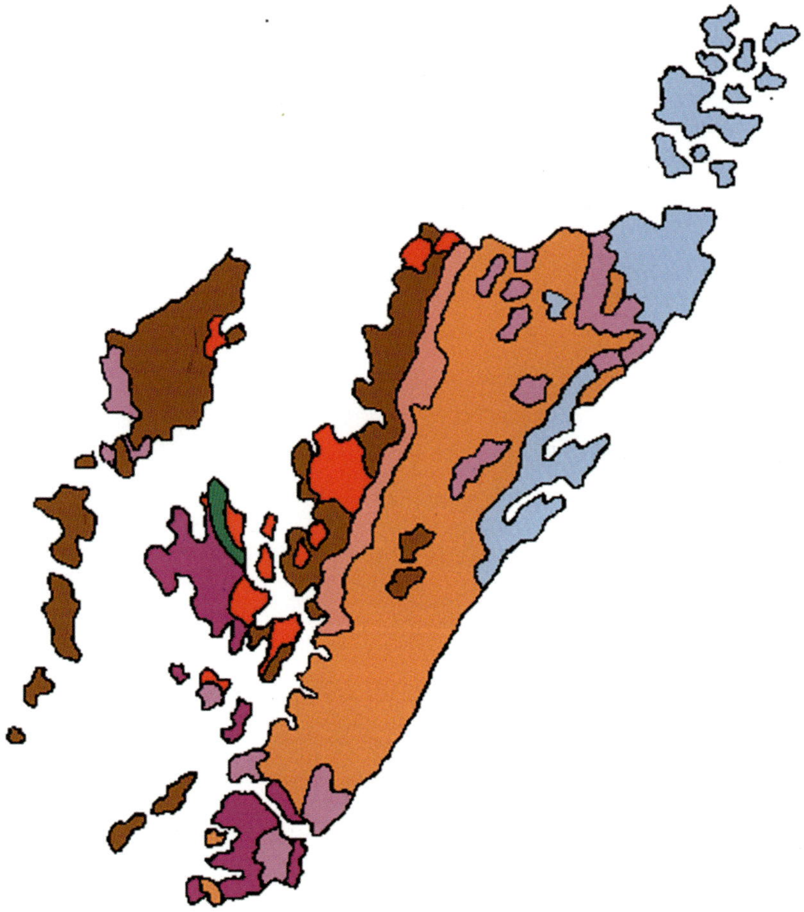

Period - Age (MYA)	*Key*
Quaternary 0-65	
Cretaceous 65-142	
Jurassic 142-205	
Triassic 205-248	
Permian 248-290	
Carboniferous 290-354	
Devonian 354-417	
Silurian 417-443	
Ordovician 443-495	

Neoproterozoic 545-1000	
Cambrian 495-545	

Metamorphic Rocks	
Palaeozoic 500-1000	
Proterozoic 1500-3000	

Igneous Rocks	
Intrusive	
Volcanic	

Scottish Highlands

Summary of Geology

The geology of the Highlands is most interesting and covers a broad spectrum from very ancient Lewisian gneisses of around three billion years of age, to much more recent sedimentary layers. The underlying rock strata continues the south-west to north-east orientation and consists primarily of rocks of the upper Proterozoic period around 500–1,000 million years of age. However, in the upper north west of the mainland and in the Isle of Lewis and throughout the Hebrides, there are significant swathes of rocks from the lower Proterozoic and Archaean periods between 1,500 and 3,000 million years of age. There are also regions of igneous and volcanic rocks to be found both on the mainland and western isles, reflecting historic volcanic activity throughout the west of Scotland. The north west of the Highlands additionally feature a thin band of rocks from the later Cambrian period around 495–545 million years ago, and there are more recent sedimentary layers to be found throughout the isles and on the mainland.

Overview

The Highlands may, in some respects, be considered an area of contrasts with a variety of sedimentary and igneous rock types to be found across landscapes which are equally varied in their own right. There is evidence of glacial activity in many areas as well as volcanic and tectonic activity, including several interesting faults. The Great Glen Fault is particularly significant, effectively dividing the Highlands into two primary regions: the Northern Highlands towards the north west and the Grampian Highlands towards the south east. The Great Glen itself runs for over 100 km from Fort William to Inverness and includes the magnificent Loch Ness, Britain's deepest freshwater loch, within its path. It is an impressive natural feature along which the equally impressive strike-slip fault runs. Indeed, the fault continues beyond the glen in the west, cutting through along Loch Linnhe and out into the sea towards Ireland. It is thought that the Great Glen Fault was active in Devonian times and again, later on in the middle Jurassic period. Relevant activity before the Devonian is harder to pinpoint but, certainly, this was and remains a dominant feature of the Highlands landscape. Smaller faults run parallel to the great fault in some areas, although their displacements are considerably less.

Another type of structural deformation, usually associated with colliding plates or rock formations, is that of thrust tectonics, which often result in older underlying strata being thrust up on top of younger rocks. Observation of such phenomena has sometimes puzzled early geologists, although we have developed a more complete understanding in recent times. A good example of thrust tectonics is provided at Knockan Crag where older rocks from the Moine Supergroup have been pushed up on top of younger rocks. This instance is part of what is now referred to as the Moine Thrust zone, a series of easterly dipping thrusts of which the Moine Thrust is the oldest and runs in a south by south east direction from the north west coast around Loch Eriboll, touching the south east coast of the Isle of Skye and the west coast of the Isle of Mull as it runs its course. There are clear examples of the older Precambrian Moine rocks overlying the younger Cambrian sequences of gritstones and limestones. It was Benjamin Peach and John Horne who first interpreted this phenomenon correctly, publishing their conclusions in a 1907 paper which has remained a reference for geologists ever since.

The north west coast also exposes some of the oldest rocks in Scotland. At Scourie there

J. Ashbourn, *Geological Landscapes of Britain*,
DOI 10.1007/978-90-481-8861-1_14, © Springer Science+Business Media B.V. 2011

exists an interesting group of metamorphic rocks which date back almost three billion years. These Precambrian rocks are usually referred to as Lewisian (from the Isle of Lewis) and are thought to underlie much of Scotland although they are exposed only in the north west and Hebridean islands. The exposed rocks at Scourie lay between the Moine Thrust and the sea and are generally grouped into the three areas of north, central and south. It is those in the central area, between Scourie and Gruinard Bay, that are particularly interesting as they seem to have not been modified since their creation more than two billion years ago. These rocks are of igneous origin and tend to be banded gneisses, although those in the north and south complexes have been subject to further metamorphism since their original creation. There are also numerous dyke intrusions in this area, generally oriented in a north-west to south-east plane, indicating a period of intense activity.

This activity is reflected in the western isles which feature a variety of volcanic and metamorphic complexes covering a remarkably broad period of time. Some of the oldest rocks in Britain are to be found here. Indeed, much of the Outer Hebrides are formed of Lewisian gneisses up to three billion years of age. These rocks of igneous origin have been metamorphosed into a variety of forms and are to be found also on Raasay, Skye, Rum and Coll. But there are also younger sedimentary rocks to be found in the region, including sandstones and limestones between 450 and 1,000 million years of age. These would have originally been laid down in shallow seas or river systems and overlain in some areas by much later deposits. In fact, Raasay, Skye, Rum and Eigg exhibit rocks from the Mesozoic era between 90 and 245 million years of age, providing an overall geological picture spanning a vast range of time. The visible and often quite beautiful landscape of these islands has

Fig. 1 Extreme stratification of the rocks towards Mallaig

also been sculpted by glacial activity, leaving a trail of ice-scoured rocks and other glacial features although certain areas, such as parts of St. Kilda, may have escaped the ravages of advancing ice. Skye and Cuillin provide good examples of glacial landscapes, while St. Kilda is a little more idiosyncratic with a varied landscape dominated by high cliffs, indeed, the highest cliffs to be found in Britain. Skye and Rum also provide evidence of lava flows and igneous intrusions, revealing a certain amount of volcanic activity. The Cuillin Hills provide a contrast to the lava flows on Skye with particularly rugged igneous formations known as the Black Cuillin and Red Cuillin, the former composed predominantly of dark gabbros and the latter of lighter coloured granites, while both show evidence of intrusive dykes. The little island of Staffa within the inner Hebrides is of entirely volcanic origin, composed largely of the distinctive columnar basalt which is clearly visible at the famous Fingal's Cave, an inspiration for many artists over the years. These distinctive lava formations were created from early lava flows from the Teritary Mull volcanic centre and are relatively silica-rich. There are also thin layers of ash and soil to be found, indicating perhaps a fairly turbulent history. Staffa is an interesting island among several interesting islands. Indeed, together, the north western isles of Scotland contain a wealth of geological textures and formations which tell us a great deal about the history of the region and its relationship with the mainland.

The northern isles Orkney and Shetland also exhibit an impressive variety of rock types and formations, as if encapsulating the geology of the Highlands in miniature. Ancient Lewisian gneisses are exposed in the Shetland isles, together with a variety of younger rocks, often in dramatic formations which sometimes seem to form an uncomfortable boundary with the sea. There are Moine and

Fig. 2 The dark rocks on the eastern shore of Loch Linnhe

Dalradian metamorphic rocks as well as Devonian igneous and sedimentary sequences. There is even an example of an ophiolite, an area of oceanic crust which has been exposed at Unst and Fetlar. This geological tapestry spans a time period from around 360 million years ago to well over a billion years ago, all contained within this small group of islands. The Orkney isles, by comparison, are much more uniform, comprised predominantly of Devonian sedimentary sandstones which were originally laid down in the Orcadian basin, an area which drained the rivers of the Highlands to the west and south. Perhaps unsurprisingly, these rocks have been found to contain some interesting and well preserved fossils. The contrast between the Orkneys and the Shetlands from an underlying geological perspective is quite interesting given their relative geographic positions. The visible landscapes have rather more in common, with areas of sandy beaches, rugged coastal rock formations and hills, although the Shetlands have rather more evidence of glacial activity, as might be expected. Both groups of islands provide something of a haven for wildlife and there are many species who either reside in, or are frequent visitors to the region.

Moving back inland to Glen Roy, not far from the Great Glen, some interesting glacial features present themselves in the form of what appear to be parallel roads cut into the landscape. They are in fact the shorelines of a great glacial lake which sat in the glen during the last glacial phase known as the Loch Lomond Stadial around 10,000 years ago, wherein glaciers re-advanced briefly through the Highlands. The periodic advance and retreat of the glacier has created three distinct levels of shoreline which have become cut into the valley sides, creating the distinctive 'roads' which are clearly visible today. These incisions are thought to have been caused through frost action, impacting upon already

Fig. 3 Soft sedimentary layer in a stream bed leading down to Loch Goil

weathered terrain. Together with other glacial features, they provide Glen Roy with a particularly interesting landscape. Loch Lomond itself is no less interesting, being created by the same glacial action which gouged its way southward through the landscape and then retreated, leaving the hollow which became the loch. As sea levels raised and lowered throughout this period, the southern end of the loch was variously invaded by the sea, leaving a deposition of marine beds and distinctive shore platforms which characterise the area.

Relative sea level is of course currently topical with a good deal of research being undertaken into the effects of isostatic rebound and associated raised shorelines. There is some interesting evidence along the Firth of Forth for example which suggests post glacial uplift as the ice from glaciers in the Forth valley retreated to the west of Edinburgh. The ice-front seemed to have stabilised for a period around Perth before a warmer period, the Windermere Interstadial, marked the banishment of significant glaciers from the Highlands. The Loch Lomond Stadial saw the return of the ice as temperatures dropped once more, with glaciers forming on Rannoch Moor and flowing down through the Loch Lomond basin, causing this part of Scotland to be once again depressed by the weight of the overbearing ice. Of course, sea level also fell during this period, only to rise again when the ice disappeared. Indeed, for a while rising sea levels took prominence over isostatic rebound with respect to relative sea level. It was as sea level rise slowed that the effects of isostatic rebound and gently raised shorelines became more developed. On the west coast of Scotland there is also clear evidence of raised shorelines, a phenomenon noticed around the turn of the last century and formally summarised by W. B. Wright in his work 'The Quaternary Ice Age', published in 1914. The overall picture is quite complex however due to the lack of precise data around

Fig. 4 A river of rocks at Glen Lyon

ice coverage and to the varying factors inherent in our understanding of relative sea level, although isostatic rebound has clearly been such a factor during the past 15,000 years.

Long before these events occurred, the momentous Caledonian Oregony brought together the land masses of Baltica, Laurentia and Avalonia, closing parts of the Iapetus Ocean and crumpling the land at points of impact. The Highland Boundary Fault which runs across Scotland is one of the features resulting from this intense activity. This fault can be clearly observed on the coast near Stonehaven where a thin layer of sediments separates the rocks of the Dalradian highlands and Devonian lowlands. Along the fault, the Highland Border Complex rocks are variously exposed and are thought to represent an ophiolite with evidence of pillow lavas and marine sediments. The Dalradian rocks are themselves typically composed of deep marine sediments which have been subsequently metamorphosed into various schists and slates, while the Devonian rocks tend to be sandstones, shales and conglomerates. The coast around Stonehaven is consequently an interesting site at which to observe evidence of this dramatic history.

The varied geology of the Highlands adopts some often quite beautiful expressions. Many of the dark volcanic rocks in the west support a variety of colourful lichen giving them a unique appearance. In the vicinity of the locks, particularly those open to the sea, a variety of flora similarly interacts with the underlying geology to create some unique landscapes. In fine weather, there is little to equal the picturesque vistas presented in the Highlands where the quality of light is also subtly different from that in lower latitudes in Britain. However, the weather is often changeable within quite short time spans in this part of the world and serves to occasionally paint a quite different picture. One appreciates this natural reality when, for example, ascending Ben Nevis. Such an

Fig. 5 The rugged hillside at Glen Croe

undertaking can often expose you to almost every type of weather within a single day. It will also serve to introduce you to a variety of geological forms and provide a vantage point from where you can really start to understand the geology of this region.

Fig. 6 Water sculpting the rocks as it flows through the glen

Fig. 7 The shores of Loch Sween

East Anglia and the Chilterns

East Anglia

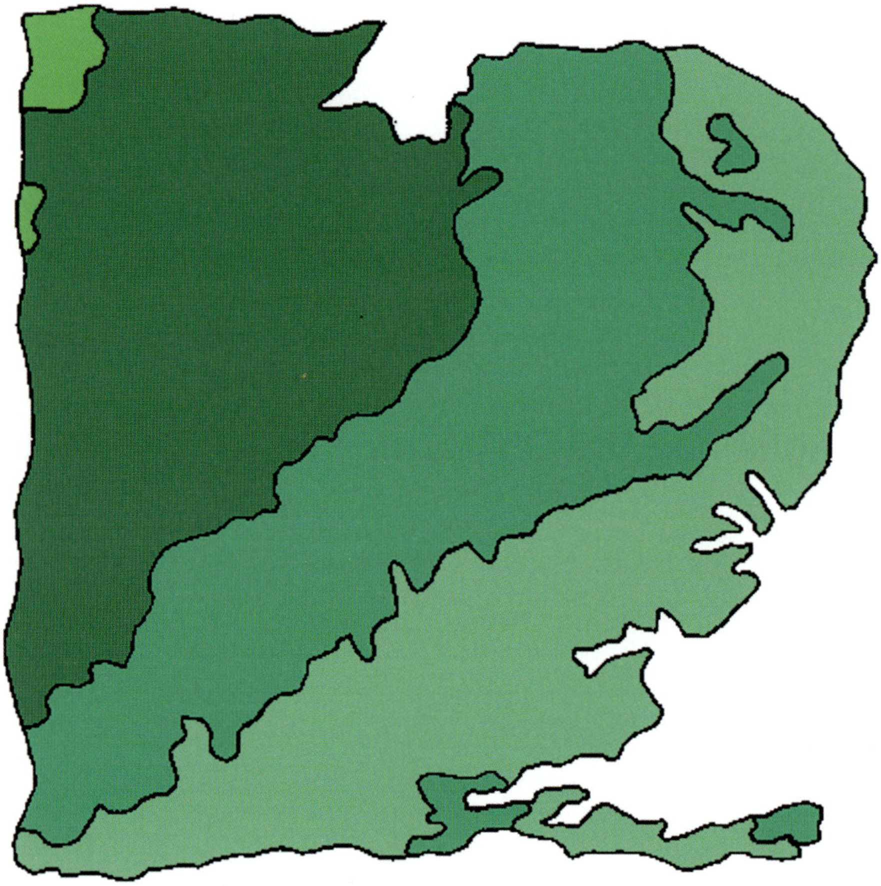

Period - Age (MYA)	Key
Quaternary 0-65	
Cretaceous 65-142	
Jurassic 142-205	
Triassic 205-248	
Permian 248-290	
Carboniferous 290-354	
Devonian 354-417	
Silurian 417-443	
Ordovician 443-495	

Neoproterozoic 545-1000	
Cambrian 495-545	

Metamorphic Rocks	
Palaeozoic 500-1000	
Proterozoic 1500-3000	

Igneous Rocks	
Intrusive	
Volcanic	

East Anglia and the Chilterns

Summary of Geology

The geology of East Anglia is quite varied with a variety mudstones, sandstones, shales and of course chalk. The chalk is visible in many areas but is often overlaid by other deposits such as boulder clay and the widespread London clay. In some areas, there are many layers of rocks, the older layers dating back around 350 million years and laying up to 300 m below ground level. However, East Anglia is also home to the youngest rocks in Britain, some strata being little more than a million years of age. The varied geology has supported equally varied habitats over the years which, in turn, have supported a variety of life, as reflected in a particularly curious fossil record which includes remnants of elephants and hippos as well as a variety of marine life. The variety of deposits and their various states of erosion makes for an interesting variation of exposed rocks and soils which can take various expressions within a given geographic area, apparently rendering East Anglia a fairly complex geological entity. However, all the visible rocks of this area are relatively young, laid down mostly in the Palaeogene and Cretaceous periods, with some Jurassic strata as one moves inland towards the lower Midlands. The area is also characterised by a tilting towards the east by south-east, a factor which may continue to raise concerns in coming years.

Overview

While the land we now know as Britain was still being shuffled around by tectonic processes, the area of East Anglia was a series of muds and sands which, around 250 million years ago, found itself overlain by a tropical sea. Further deposits of mudstones, sandstones and limestones where laid down followed by chalk. Then, from around 55 million years ago, the area was raised above sea level and, over many millions of years, systematically tilted downwards toward the east. As erosion took place, aided by the ravages of cyclical climate change, the edges of various sedimentary layers were exposed at the surface, creating the rather mixed geological landscape we see today.

A notable event in the geological history of East Anglia was the Anglian Glaciation which took place around 500,000 years ago and covered the whole of the area in a thick ice sheet. Subsequent periods of warming and cooling, the last being the glaciation around 18,000 years ago, have been accompanied by rising sea level, causing the eastern extremities of East Anglia to flood and be swamped with mud and peat. Such landscapes characterise the fens and the Norfolk broads. Elsewhere, such as across the Chilterns, the chalk is exposed in many areas while clays and gravels can be found almost everywhere. Actually, the chalk is never far away and is expressed in one form or another from Dover and Seven Sisters on the south coast, right up to Hunstanton on the north Norfolk coast, where it takes on a curious pink colour in places.

To the west of East Anglia, the county of Hertfordshire presents an interesting picture with the gently rolling hills and valleys of the Chilterns concealing an underlying geology of Cretaceous chalk, overlain in many areas by clays and gravels. In the south east of the county, the London clay of the Thames basin is very much in evidence with smaller areas of gault clay to be found in the north west and boulder clay in the central and eastern areas. These clay deposits often contain scatterings of flint, particularly in the West Hertfordshire area where flint nodules of various sizes are often exposed at the surface. The Chilterns chalk tends to present a steep north westerly facing scarp with more gentle south easterly slopes rolling down towards the Thames Basin

J. Ashbourn, *Geological Landscapes of Britain*,
DOI 10.1007/978-90-481-8861-1_15, © Springer Science+Business Media B.V. 2011

area. The chalk is exposed in several areas, notably on the Dunstable Downs where the topography described may easily be observed, with a distinct north by north westerly facing scarp with more gentle slopes to the south. The higher points of this chalk scarp are just over 240 m in height and provide a commanding view in all directions. Away from the slopes of the Chilterns, the Hertfordshire landscape becomes somewhat flatter with mild river valleys wherein the clay deposits are often interspersed with gravels.

Moving eastward to Essex, there is a not dissimilar picture of underlying bedrocks overlain by various sedimentary layers. Essex borders the Thames estuary and is part of the Thames Basin which features a good deal of London clay at the visible level and substantial deposits of boulder clay towards the north west of the county. The bedrocks of Essex consist of shales mudstones and sandstones originally laid down around 350 million years ago. These rocks are now around 300 m beneath the surface however, overlain by a series of deposits. These include a significant layer of gault clay and sandy deposits laid down around 250 million years ago, followed by the Cretaceous chalk which underlies the Thames Basin and becomes visible in just a few areas. On top of the chalk are various layers of sands and clays, including around 50 million years ago, the laying down of the London clay within the Palaeogene period. Further deposits ensued and the landscape was sculpted somewhat by the glaciations of the last ice age which served to slightly alter the course of the Thames and other rivers, leaving a good deal of exposed gravel and sandy deposits, as variously reflected within the Thames estuary. Essex consequently provides a mix of mostly clays and gravels at the exposed level with an interesting stratification beneath, reflecting an equally interesting history of sedimentary layering.

Fig. 1 The end of the Chilterns at Dunstable

To the north of Essex, the Suffolk geology presents a fairly straightforward mix of clays, gravels and sands overlaying a layer of chalk which is rarely exposed at the surface, but is occasionally visible in the west. Much of the landscape has been sculpted by ice age glaciations and fluctuating sea levels resulting in the extensive boulder clay deposits which characterise much of Suffolk, particularly in the central area. Such deposits often contain fragments of chalk and flint which have been scoured from the underlying rocks by glacial activity. Towards the east, gravels and sand play a more dominant role, with marine sandstones (sometimes referred to as The Crag) exposed on the coast. These rocks are often fossil rich and provide further evidence of the part played by rising and falling sea levels within the creation of this landscape. Suffolk reflects the general East Anglian tilt towards the east, whereby the underlying chalk has a wedge of subsequent deposits, becoming thicker towards the eastern coastline. The variance of clays, gravels and sands (and occasionally chalk) exposed at the surface is reflected in the natural habitats which include woodlands atop the central clays and heathlands atop the eastern gravels and sands. The coastline itself includes low cliffs and some interesting pebble beaches, some of which have played an important part in the history of the county. Today, some of these areas support nature reserves, aided by the several estuaries which characterise the Suffolk coastline. Woodlands still play a significant part in this respect and are evident throughout Suffolk.

Moving further north, Norfolk has a certain amount in common with Suffolk from a geology and landscape perspective, including the Crags that punctuate the coastline and a general picture of more recent deposits towards the east, overlaying a series of clays and a layer of chalk. The Norfolk chalk is generally acknowledged to be among the youngest in

Fig. 2 Typical flat East Anglian fens

Europe, laid down between 65 and 100 million years ago. In parts of the county there exist pockets of an older and unusual red chalk laid down around 105–125 million years ago, as can be observed within the cliffs at Hunstanton where a delightful visual contrast may be found between the lower strata of the Hunstanton Red Chalk, overlain by the pale creamy colours of the Ferriby Chalk Formation. The colour of the red chalk is considered due to the presence of limonite, an ore rich in iron oxide and often associated with low rates of sedimentation. Flint nodules from the chalk are to be found in relative abundance across the county and have provided an important building material over the years, as reflected in some of the older buildings. The subtropical marine environments that existed throughout these periods are reflected in the fossils among the crags, mudstones and sandstones throughout East Anglia and particularly in Norfolk which has become an important site in this respect from a global heritage perspective.

The Norfolk Broads provide an important haven for flora and fauna within the largest protected wetlands in Britain and are recognised internationally as a focus point for conservation. On the north Norfolk coast, particularly around Cley and Blakeney, the wetlands also provide a natural habitat where wildlife may thrive, especially the various birds and waders that frequent this area, including spoonbills which seem to be making something of a comeback. The wetlands are additionally enhanced by a number of widely scattered natural ponds which have been named 'pingos' after the Inuit word for small hills. This reflects their ice age heritage whereby hills of ice subsequently collapsed in the warmer weather, creating depressions filled with water. These pingos are consequently many thousands of years old and there are many of them scattered about Norfolk

Fig. 3 The gentle slopes of a typical Hertfordshire landscape

and, indeed various other parts of the country. This combination of ponds, marshes and wetlands is enhanced by salt-marshes in areas where natural lagoons form behind shingle ridges such as those around Blakeney, thus supporting a wide variety of wildlife.

The Norfolk coastline is mainly characterised by a soft fine sand, creating grassy dunes in places and abutting to often quite shallow sandstone and mudstone cliffs. Consequently it is easily eroded and several areas along the coast are experiencing problems in this respect, with cliffs disintegrating and falling down onto the shoreline where fragments are easily dissolved and washed away into the sea. In other areas, mudslides have caused the cliffs to adopt and almost step-like character, often creating basin like depressions at the coast line as erosion continues. This is often caused by the ponding of groundwater due to more or less impervious layers which, in turn, can cause slippage and associated landslides. Indeed, a variety of cliff formations may be experienced within fairly short distances depending upon the precise exposed strata and its relative permeability. As one moves northward along the coast, the cliffs reflect a subtly different range of visible compositions with clear sedimentary banding in evidence beyond Mundesely, sometimes exposed by collapsing sections and mudslides. A remarkable collage of yellows, browns and greys is often to be found along this part of the coast.

East of Norfolk and into Cambridgeshire, the landscape retains a fairly flat composition across the Cambridgeshire fens. Much of this area was laid down between 140 and 200 million years ago during the Jurassic period, as was much of Northamptonshire. Indeed, this swathe of Jurassic strata reaches down to the Dorset coast and up to the north Yorkshire moors. To the south and east of this swathe lay the younger Cretaceous deposits reaching down to the south coast, with yet younger

Fig. 4 A shingle beach at Orford in Suffolk

Palaeogene deposits around the Thames basin and along the Suffolk and Norfolk coasts.

East Anglia, while not perhaps featuring quite as dramatic a geology as may be found elsewhere in Britain is nevertheless important from a geological perspective, providing an interesting example of relatively recent sedimentary strata superimposed upon rocks which themselves are fairly young in geological terms. The landscape and supported environments reflect the comparatively dynamic situation of rising and falling sea levels and glaciations which have served to shape the region overall and the eastern coastal areas in particular. Human activities have additionally served to accentuate some of these processes: for example the exploitation of peat in the Norfolk Broads area and dredging for aggregates off the coast. Indeed, the East Anglian coast continues to be subject to change with coastal erosion becoming a significant issue in certain areas. An example of geological and related natural processes in action.

Fig. 5 The chalk is never far away in East Anglia

Fig. 6 The soft sandy coastline at Winterton in Norfolk

Fig. 7 The East Anglian geology supports many woodlands

Coastal Erosion

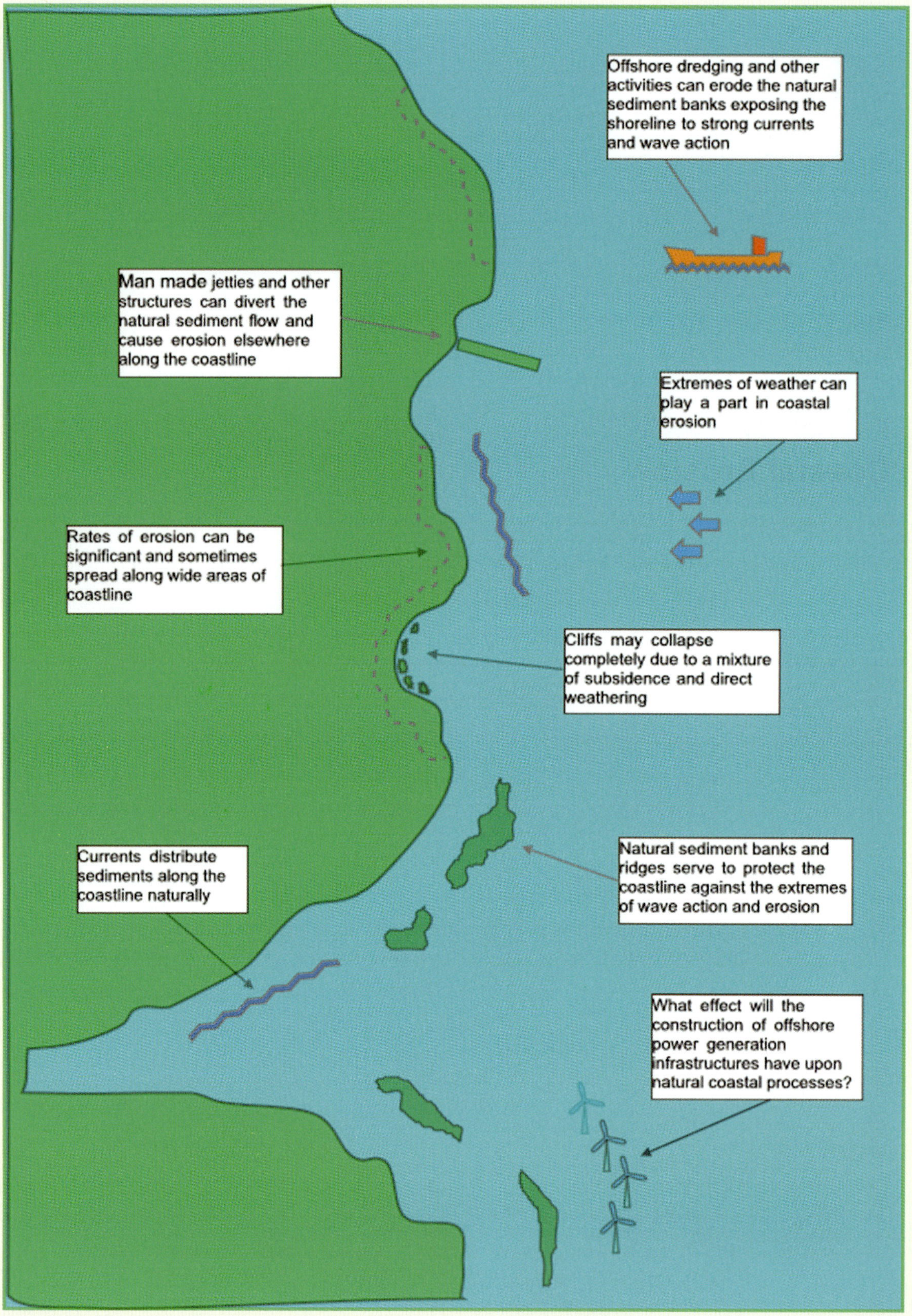
Offshore dredging and other activities can erode the natural sediment banks exposing the shoreline to strong currents and wave action
Man made jetties and other structures can divert the natural sediment flow and cause erosion elsewhere along the coastline
Extremes of weather can play a part in coastal erosion
Rates of erosion can be significant and sometimes spread along wide areas of coastline
Cliffs may collapse completely due to a mixture of subsidence and direct weathering
Currents distribute sediments along the coastline naturally
Natural sediment banks and ridges serve to protect the coastline against the extremes of wave action and erosion
What effect will the construction of offshore power generation infrastructures have upon natural coastal processes?

Coastal Erosion

Summary

When we hear the term 'coastal erosion' we tend to think of extreme processes eating away significant portions of coastline within a very short time-frame. Certainly, such extreme conditions do exist but we should perhaps remember that coastlines are never completely static and stable. Where any land mass meets the sea, there is bound to be some sort of interaction with tides, winds, discharge from rivers and other natural factors which, in turn will have some sort of effect upon the coastline itself.

Sediments are usually in a state of flux, being either deposited or eroded at various places along the coastline, depending upon a number of factors including tides and longshore currents, wind and weather and other natural processes. Furthermore, such sediments may range in size from tiny grains of sand and natural water borne sediments to quite large pebbles and, occasionally, large chunks of material broken loose by weathering. Furthermore, it is likely that we shall be subject to rising sea levels due to glacial melting as we continue our exodus from the last glacial period, further exacerbating the issue.

In addition to these natural processes there are man-made causal factors, often brought into play unwittingly due to the construction of jetties, piers, artificial harbours and re-claimed land. Any such man-made construction at the interface of land and sea will certainly have some effect. This effect may be marginal, or it may be catastrophic. The former may simply re-shape the coastline somewhat, causing sediment to be re-distributed in an unexpected

Fig. 1 The soft mud and sandstone cliffs on the Norfolk coast

J. Ashbourn, *Geological Landscapes of Britain*,
DOI 10.1007/978-90-481-8861-1_16, © Springer Science+Business Media B.V. 2011

manner. The latter may result in serious flooding and attendant damage to both natural and mad-made environments, even causing loss of life. Human activities off-shore, for example aggregate dredging, the erection of power generating constructions etc., can also play a significant part in erosion as natural processes are interrupted.

Our knowledge of natural processes and the likely effect of disturbing them should render us better prepared to evaluate proposed constructions and advise accordingly. Similarly, proposed responses to coastal erosion which involve yet more man-made devices need to be evaluated very carefully to ensure that we are not simply exchanging one problem for another. We shall need to consider the island as a whole, including all of its coastal processes, if we are to devise an intelligent and sustainable approach to managing coastal erosion. In any event, coastal erosion is an enduring issue for many island communities, including Britain. In this section of the book, we shall consider some examples of this phenomenon.

Overview

Currently, there are several areas of the British coast that are giving cause for concern due to perceived coastal erosion, although the issue is of particular pertinence to the south and east coasts. Such is the situation that the Environment Agency is producing projections as to which areas of coast are likely to be most affected during the next 25, 50 and 100 years in order to prioritise strategies for defensive action. In parallel, research teams have initiated sophisticated mapping projects in order to accurately quantify the severity of the problem.

The dangers of coastal erosion have been understood in Britain since medieval times when the village of Dunwich, once a thriving centre of wool trade, was systematically eroded and eventually disappeared. A more

Fig. 2 Shallow cliffs sculpted by the continuous action of the sea

recent and more extreme case was the coastal village of Hallsands in Devon which was quite literally washed away overnight. Depending upon the precise nature of the coastline, and how human habitats have been constructed in proximity to it, erosion can easily escalate to such extremes. Currently, the area of Holderness in the north-east of Britain is considered particularly vulnerable. This is partly due to its stratigraphic composition of relatively young sedimentary deposits of mud sand and gravel (around 12,000 years of age) left behind by the last glaciation and sitting atop a layer of chalk which itself is sloping gently to the east. Indeed, well before the glaciation, around 30,000 years ago, the whole area of Holderness would have been under the North Sea. As the North Sea ice melted, leaving the aforementioned deposits, the coastline adopted a profile which would be familiar to us today, although erosion continued to be an issue. In more recent times, much of the surrounding marshland has been drained and a certain amount of land reclaimed. However, the spectre of erosion has never really gone away and Holderness continues to be buffeted by the waves of the North Sea clawing at the relatively soft muds and clays of the fairly steep cliffs. Such action causes the cliffs to be weakened and sometimes quite substantial chunks of the cliff to be detached. Initially, the resultant debris may actually form a guard against erosion as the effective cliff is now less steep. However, longshore drift action removes much of the debris to points further down the coastline and allows the cycle to repeat.

There is historic evidence to suggest that towns and settlements in the Holderness area, many of which grew as Victorian resorts, have been affected over time. In 1546, there is written evidence of finding decayed houses (due to the encroaching sea) at Hornsea Beck, together with decayed ground in the fields to the extent

Fig. 3 Defences such as these are proving inadequate

of '12 score yards'. Later at Hornsea, we find that the measurement to the sea between the years of 1759 and 1836 had diminished by 133 yards. At Withernsea, it is suggested that coastal erosion between the years of 1852 and 1882 averaged 2.3 yards per year and that between 1845 and 1911 a strip of 98 yards wide had been washed away. At Kilnsea in 1826, the village church collapsed due to subsidence caused by erosion. Defences in the form of sea walls and groynes have variously been built in the area, such as the groyne at Mappleton and the sea wall at Withernsea for example, and these have had some effect, albeit sometimes at the expense of adjacent areas as natural distribution patterns are interrupted.

At Happisburgh in Norfolk, erosion continues to be an issue, in spite of various makeshift obstacles being placed in the path of the encroaching sea. The relatively soft, sandy cliffs are easily eroded by fierce wave action and this is now posing a threat to buildings in proximity to the coastline. Happisburgh, with a population of around 1,400 contains several listed buildings, an interesting lighthouse and a lovely old stone church. The village used to be some distance from the sea but some buildings are now perilously close to the edge of the cliffs. Between the years of 1,600 and 1,850, over 250 m of land was lost to the sea and more recently, in 1953, there were heavy floods in the area which claimed 76 lives. The wooden defences built subsequently to that time are now in a state of disrepair and are becoming increasingly less effective as Happisburgh is once more struggling with the spectre of erosion, with the coastline changing visibly year by year. Interestingly, in places this erosion causes the cliffs to adopt a step-like profile due to the relative hardness of the different sedimentary layers. This is readily seen in areas where a lower mudstone layer has proved more resilient than a higher

Fig. 4 Cliff collapse is common along sections of the North Norfolk coast

sandstone or soft sand and mud layer. Such a sculpted landscape can act in some ways to redirect the water along the coastline until it finds a softer, less resistant cliff face.

As one heads further along towards the north, the coastline between Mundesley and Trimmington displays some extraordinary formations of broken cliffs and mudslides wherein one can plainly observe the various sedimentary layers and how these are being eroded by the relentless lapping of the North Sea tides, in some cases causing collapses further back as semi dish-like areas collapse and slide down toward the waves. These areas can become quite large, forming bay like indentations into the natural line of the cliffs. On the top of the cliffs, once well trodden footpaths often seem to disappear over the edge of the cliff as the land on which they were founded is simply no longer there, having collapsed and, in many cases, been washed away by the sea. This whole stretch of Norfolk coastline, from below Happisburgh up to Sidestrand and Overstrand has various issues with erosion of one sort or another. In places, defences and sea walls are containing the issue, in other places, no such defences exist. Erosion of cliffs at the coastline is often accompanied by landslides and subsidence which effectively extend the problem inland of the waterline. This has been an issue at Lyme Regis on the south coast, an area famous for its fossils where many of the fossil rich regions have now been effectively covered by mudslides, which still seem to be on the move. The cliffs to the east of the town seem to be especially vulnerable with a good deal of collapse having taken place in recent years. A similar picture exists at Albrough on the Yorkshire coast where large chunks of the cliff have fallen away over the years, taking with them parts of the road and more than one hotel. A varied stratigraphy at Albrough renders parts of the cliff-face more vulnerable to erosion

Fig. 5 An unstable cliff at St Govans in Pembrokeshire

according to height. This in turn tends to produce a step like profile which lends itself to further landslides and subsidence. Between 1852 and 1951 annual rates of erosion at Albrough, derived from local records, seem to have been around 1.2 m. Between 1951 and 2004, this rate had increased to around 2.2 m per year, a significant increase. Similarly, further down the coast at Sidestrand in Norfolk, significant landslides and mudslides have played their part in coastal erosion. Some of the landslides in the Sidestrand area have been fairly deep-rooted and expose layers of clays and tills which, interestingly, show significant folding by previous tectonic processes.

Research at several such sites is unveiling an interesting picture and helping to produce an accurate map of how erosion is changing the British coastline. The issue, interesting though it may be from a geological and geographical perspective, has also become political as government agencies struggle with the potential cost of defending these coastal areas and have suggested that some areas should simply be abandoned altogether. Such suggestions are of course bound to be controversial. However, with the rates of erosion currently being witnessed, it is clear that important decisions will have to be made in the foreseeable future. Currently, there is much concern around government plans to abandon rural areas and communities to their fate and focus on relocation rather than defences. Villages such as Overstrand in Norfolk, Leysdown on Sea in Kent and Bawdsey in Suffolk have apparently already been earmarked for destruction, while others will become increasingly vulnerable to serious flooding. Such a policy is in stark contrast to the situation in The Netherlands, where the coastline is rigorously defended and affected communities in Britain will naturally ask why the same policy should not prevail here. Consequently, the next few decades will be significant for the

Fig. 6 Mudslides at Lyme Regis have claimed several beach areas

British coastline, particularly in areas such as East Anglia and the south coast. If current governmental thinking prevails, it may mean re-drawing the map of Britain. There will also be serious implications for the ongoing availability of arable land in affected areas. The Environment Agency will now take responsibility for the overall coastal perspective and associated sea flooding risk management. This agency will work with local authorities who will manage the risk of coastal erosion at the regional level and strategic coastal groups will be created as appropriate. However, much of this hierarchy seems to be mostly concerned with the distribution of funds according to the submission of specific project proposals. The coordination between such proposals and indeed, the longer term strategic view of how such proposals might work positively together is another matter. Indeed, there may be risks that an inappropriate short term proposal in one area may cause more damage than it prevents when viewed from a more general perspective. Perhaps we need a good deal more scientific analysis of causes and effects around the entire British coastline in order to conceive a longer term sustainable strategy.

Another issue which has come to light on the south coast is that of erosion exposing landfill sites and thus raising the possibility of serious pollution, including possible toxic pollution, depending upon what has been dumped in these sites. In addition, the prospect of beaches being strewn with decaying rubbish is hardly an attractive one for the local authorities concerned, especially as many such regions rely on tourism for part of their overall economy. Such issues serve to highlight the reality that there is always the unexpected and that we must think very carefully about areas affected by coastal erosion and what the real effects of continued erosion might actually mean. There is also the

Fig. 7 Unstable cliffs on the Lleyn Peninsula

possibility of salt water encroaching upon pockets of fresh water and the effect that this might have upon natural environments and land use. Indeed, coastal erosion is not just a matter of losing a few metres of cliffs, there are a range of possible knock-on effects which should be understood and taken into consideration before abandoning whole areas of coastline. Such matters should also feed into the broader strategic view.

The situation may be further exacerbated by rising sea levels as we continue to exit from the last period of glaciation, although this is somewhat difficult to quantify with any accuracy. Throughout the next century, this may place additional pressure upon currently affected areas and may well bring previously unaffected areas of coastline into question. Another related variable is that of weather patterns and the possibility of more intense winds and storms. If such a trend develops, then the rate of erosion at coastlines may well accelerate. If the response policy in certain areas is one of abandonment, how long can such a policy be maintained and over how many periods of serious erosion before the problem escalates to the scale of a national crisis? How much more land are we prepared to loose? The question becomes interesting from a societal perspective and is perhaps one that only history will answer.

Of course, as with many natural processes, erosion caused predominantly by natural weathering can be cyclical and there are instances where the coastline has effectively moved backwards and forwards over time. Such examples were witnessed around the turn of the last century on Merseyside at Formby point and along the Sefton coastline. In this example, between 1845 and 1975, the relatively soft shoreline initially moved seaward, in some areas up to around 300 m, before stabilising and then retreating again, the turning point for many areas being around 1920. The sites at which

Fig. 8 Collapsing cliffs on the Norfolk coast

such activity is most pronounced may also change slightly, moving along the coastline according to the precise nature of the erosional forces. This is understandable when one considers the overall sediment budget and the possibility of sediment being gained from offshore banks as well as lost from the shoreline as a result of shifting currents. While such cyclical activity is not unknown, in general the processes seem to be more one-sided with erosion outstripping progradation according to many factors including the availability of sediment and weather patterns.

In conclusion, island coastlines are always subject to movement due to the continual distribution of sands and sediments by longshore drifts and other natural processes which often maintain a sort of equilibrium from season to season. Over geological timescales, events such as glaciations and subsequent ice melts will naturally have a larger effect due to rising and falling sea levels and, over time, even these more extreme natural processes may find something approaching equilibrium although, ultimately we also have tectonic effects to contend with including the tilting of Britain towards the south east and of course movement of the tectonic plates themselves. Over shorter timescales however, one might argue that human activity is at least partly responsible for the accelerated rate of coastal erosion currently witnessed in Britain. Relatively uncontrolled large scale commercial aggregate dredging off the Norfolk coast will undoubtedly have had a major effect in this context and may even be the primary cause of the erosion we are now witnessing. In other areas where a natural offshore sediment ridge exists, such as further south at Thorpeness on the Suffolk coast, the shoreline remains comparatively stable. However, if such natural offshore ridges are removed, then the shoreline is exposed to the full force of wave action and, on soft shorelines such as those of East Anglia, erosion will surely follow. Ill-considered artificial constructions along the coastline may also play a part in interrupting the natural flow of sediments around the British coast and thus causing accentuated erosion at specific points. This may have been a factor along the south coast, where certain areas have been particularly prone to erosion in recent years. Our coastlines are in fact relatively fragile environments which may easily be adversely affected by ill-considered human activities as well as natural processes. If this book is re-written a hundred years from now, it might present a subtly different picture of the British coastline, and a hundred years is no time at all in geological terms. A thousand years hence, or ten thousand years, or even a hundred thousand years – still insignificant timescales from a geological perspective, and Britain herself may be quite different. Such is the nature of geology and natural processes over the longer term. In the shorter term however, we must strive to understand coastal erosion around Britain, its underlying causes and what we might do to manage the situation in order to best preserve our natural coastlines and attendant habitats.

Fig. 9 The typically soft composition of the cliffs at Happisburgh

Epilogue

Epilogue

In this book we have explored the geology of Britain and in so doing created an up to date photographic record of this geology and its supported landscapes. It has certainly been an interesting experience, travelling around the island and witnessing at first hand both the underlying geology and the various environments attached to it. There is a great deal of diversity of course, with fairly late sediments contrasting with, and sometimes overlaying, much earlier rocks of both igneous and metamorphic origin. When we gaze upon such scenes, we are reminded of the tectonic and related processes which have lead to the creation of Britain and the British Isles, especially evident in the complex folding and tilting of many visible rock layers, both at the coastline and inland. While this complex construction was orchestrated over many hundreds of millions of years, with numerous fundamental events undertaken at various points across the globe, it nonetheless gives the appearance of permanence, at least in the context of human timescales. But this, of course, is an illusion. Nothing is permanent. The tectonic ballet will continue and new lands will be created while others are subsumed back into the melting pot. It is the natural way of things and will remain so, just as long as Planet Earth retains its life-giving heartbeat. In time, the area of land we know as Britain will change again, possibly into something quite different from what we would recognise today, as a result of these natural processes.

In Britain, we are witnessing some of these processes upon a much finer scale in the form of coastal erosion. The phenomenon is not

Fig. 1 Folded sedimentary layers at Broadhaven

J. Ashbourn, *Geological Landscapes of Britain*,
DOI 10.1007/978-90-481-8861-1, © Springer Science+Business Media B.V. 2011

unique to Britain of course. France has similar problems along its northern coastline as do many other countries whose lands kiss the sea. Ultimately, there may be little enough we can do to alter the progress of such elemental processes. However, we may be able to lessen their impact within the shorter term with a little intelligent thought and planning. Unfortunately, we often seem to be accentuating such issues via careless human activities, usually commercially inspired, which accelerate the damage to this delicate natural fabric. Within these pages, you have seen evidence of coastal erosion in locations which, if we had photographed them just 50 years ago, would have looked quite different. It is an issue which we must strive to properly understand and consequently address with intelligent planning and associated action if we are to lessen the impact for future generations.

Indeed, care of our natural environment is important from many perspectives. Such a duty of care should encompass many factors in order to understand the broader picture and the complex relationships and dependencies within the natural world. Some of these factors may be non-obvious, for example soil erosion and the overall quality of soil is a vitally important factor within this broader mechanism and yet rarely is brought to our attention. Similarly important is the preservation of natural habitats, not just for aesthetic considerations, but in relation to this broader balance and the associated mechanisms. Within this broader picture, any unnecessary damage sustained in one area will almost certainly have follow on consequences elsewhere. This reality extends to wildlife as well as habitats, all of which are interlinked in a myriad of complex ways.

If future generations are to form a better understanding of such matters and better preserve their world, then an understanding of the Earth Sciences will become increasingly

Fig. 2 Sedimentary strata rising from the sea at St. Annes

important. Geology is a fundamental building block of such an understanding. Consequently, it is hoped that books such as The Geology of Britain will inspire young students especially to take a closer look at the Earth Sciences and perhaps continue with their own research in order to further develop this understanding. After all the study of geology, in addition to being an important factor in the understanding and preservation of our natural heritage, is an endeavour which can be enormously stimulating. Those who have already acquired such an understanding may find the photographs in this book especially valuable as an up to date record of the many geological forms and landscapes which characterise Britain. Others may simply enjoy the photographs for the beauty of the natural world which they encapsulate. In any event, I sincerely hope that all readers have enjoyed this book and that it may have inspired an enduring passion for Britain and its underlying geology.

This geology, as we have witnessed within the pages of this book, is at times beautiful, dramatic, fragile and complex. It is always extremely interesting and, in many ways, provides a microcosm of geology upon the broader global scale. We have much evidence of volcanic activity and the distinctive rock formations arising from it. We have wonderful examples of granite intrusions creating a variety of granite textures and even colours. We have areas characterised by compacted slate of various densities. We have coalfields and seams of coal at various locations. And of course we have other sedimentary rocks, laid down many millions of years ago in a variety of oceanic conditions, creating a wealth of textures, densities and hues from chalk to beautiful sandstones, limestones and a variety of mudstones, clays and shales. We have landscapes shaped by glacial action, with U shaped valleys, glacial tarns and other features left behind by the retreating ice and we have various features sculpted by the meltwaters

Fig. 3 Vertical stratification in the sandstone at Freshwater

that followed. Our limestone pavements and caverns create a wonder-world of their own, with fantastic patterns and structures which could only come from nature. Rich mineral seams at many locations in Britain have been appreciated and mined for hundreds and, in some cases, thousands of years. Indeed, the geology of Britain as an island provides something of a cornucopia of interesting history and associated structures to explore. This book has simply provided a taste from this cornucopia as it would be almost impossible to capture the complete story, with all its perspectives and human interest, within a single volume. Hopefully, it has inspired the reader to more deeply contemplate geology and the wonder of our physical world. If all this can be experienced within a small island such as Britain, just think of the many wonders locked within our planet as a whole, many of which may yet be fully appreciated.

If we further extrapolate our thinking beyond the realms of pure geology to that of what we might describe as 'Earth Systems', we may better place geology and geological cycles in context. Earth Systems may be defined as the complex mechanism of interactions between the lithosphere, hydrosphere, biosphere and atmosphere and the transfer of resources between them. The cyclical nature of each of these in isolation has long been understood. We understand the hydrological cycle and the links between precipitation, evaporation and the transportation and storage of natural water sources. Similarly, we understand rock cycles, tectonic processes and weathering, as do we understand thermal gradients and weather systems as well as natural cycles within the biosphere. Each of these cycles is concerned in one way or another with passing resources from one entity to another in a continuous model of recycling. We are now beginning to

Fig. 4 The sediments collide at Mallaig

appreciate (or perhaps re-appreciate) that these cycles are themselves interlinked in sometimes quite complex ways. It is the combination of these cycles with their various inter-dependencies and associations which constitute our living planet.

In many respects, we may consider the Earth as a closed system due to this continual cycling of finite resources. The exception of course is the radiated energy received from the sun which continues to act as a power source to help drive these other mechanisms. When we think of the Earth in this way, as a living, dynamic and holistic system, we can start to appreciate that almost everything is interlinked and dependent in one way or another. Life in the biosphere could not continue without the atmosphere, hydrosphere and lithosphere, each of which are interdependent. It follows then that significant change in one of these areas will likely have consequences for the others. Which brings us back to our point about care for the environment. We must appreciate that actions have consequences, some of which we may not readily understand, or may not otherwise be readily apparent.

Related to the concept of Earth Systems is the idea that the holistic system is somehow self regulating and maintains a balance by reacting to changing conditions via feedback mechanisms. This may indeed be the case. The question is, within what limits? Do we really understand the extremes of change that the Earth may be able to tolerate within this self regulation? We do to some extent, courtesy of the history held within the rocks. We certainly understand the concept of ice ages and periods of glaciation, just as we understand tectonic processes. We also understand that there have been periods of mass extinctions from a species perspective. These events

Fig. 5 Interesting colours exposed in the fragile rock at Loch Sween

alone should highlight the reality of interconnected systems. Perhaps future generations will re-define the study of Earth Sciences in order to accommodate the concept of Earth Systems and place an emphasis upon an understanding of the complex relationships and dependencies that the phrase suggests. We may have much to learn in this respect. In any event, the science of geology as currently understood will play a key part within that broader understanding. And that brings us back full circle to the geology of Britain and how the study of the same can demonstrate the reality of these broader processes upon a localised scale. It is perhaps an ideal starting point for research into Earth Systems in a broader sense.

Lastly, may I thank you for reading this book. I do hope that you have enjoyed it, not just for its discussion of geology, but in its realisation of the natural beauty inherent in geology and its many processes. It is both the essence and mechanism of our beautiful planet.

Julian Ashbourn

Fig. 6 Large sedimentary strata at Loch Claunie

Fig. 7 Dark sediments exposed at Fan Fawr

Fig. 8 Natural patterns within the Kimmeridge clay and mudstones

Fig. 9 Limestone layers at Lyme Regis

Fig. 10 History in the sediments

Appendix Figures

Fig. 1 British weather over Loch Sween

Fig. 2 Rocks folded like paper at Hartland Quay

J. Ashbourn, *Geological Landscapes of Britain*,
DOI 10.1007/978-90-481-8861-1, © Springer Science+Business Media B.V. 2011

Fig. 3 Limestone pavements at Ingleton

Fig. 4 Red sandstone cliffs at Sidmouth

Fig. 5 Granite tors on Dartmoor

Fig. 6 Chalk in the sea at Seven Sisters

Fig. 7 The vulnerable beach and low lying dunes at Caister in Norfolk

Fig. 8 The beach reclaims what was once higher ground at Caister

Fig. 9 A salt water lagoon protected by a shingle spit at Cley in north Norfolk

Fig. 10 Sections of the cliff wearing at different rates according to stratification at Happisburgh

Fig. 11 The relatively soft mudstone of the Happisburgh cliffs

Fig. 12 Interesting stratification among the sandstone and mudstone at Happisburgh

Fig. 13 The land meets the sea in a battle for the shoreline on the Norfolk coast

Fig. 14 A limestone terrace on the Black Mountain in South Wales

Fig. 15 Calcite nodules within the limestone on Black Mountain

Fig. 16 A peak of weathered limestone on Black Mountain

Fig. 17 Limestone weathering in action

Fig. 18 Near vertical stratification under the stream bed at Capel Gwynfe in Wales

Fig. 19 Run off from the hills winds its way down to the river at Capel Gwynfe

Fig. 20 A small lake among the Norfolk flat lands

Fig. 21 On the high ground above the Gower in South Wales

Fig. 22 Limestone smoothed by the waves on the Gower in South Wales

Fig. 23 Note the different thicknesses of stratification among these limestone layers

Fig. 24 Rising up from the sea, the edge of South Wales at the Gower

Fig. 25 Low dunes approach the sea on the edges of the Gower

Fig. 26 Distinct layering in the rocks at Barafundle in Pembrokeshire

Fig. 27 Interesting patterns in the rocks at Stackpole on the Pembroke Coast

Fig. 28 Folded rocks on the coast at Broadhaven in Pembrokeshire

Fig. 29 Nice even stratification among these layers at Whitesand in Pembrokeshire

Fig. 30 Colourful weathering of the rocks at Littlehaven on the Pembroke coast

Fig. 31 Lichen keep a hold high up on Ben Nevis

Fig. 32 The aged and rugged rocks on the slopes of Ben Nevis

Fig. 33 Looking out across Glen Garry in the Highlands

Fig. 34 A typical highlands landscape at Glen Lyon

Fig. 35 Boulders washed down from the hills at Glen Lyon

Fig. 36 The hills framing Glenfinnan in the Highlands

Fig. 37 Looking out across Jura Sound to the Isles of Jura

Fig. 38 Very ancient rocks adjoining Jura Sound

Fig. 39 Calcite trapped within these ancient rocks on the edge of Jura Sound

Fig. 40 Limestone boulders aside Loch Claunie

Fig. 41 The distinctive shores of Loch Linnhe

Fig. 42 The gentle hills adjoining Loch Lomond

Fig. 43 Looking north along Loch Ness and the Great Glen

Fig. 44 An interesting section of the southern shore of Loch Ness

Fig. 45 Interesting weathering among the rocks on the shores of Loch Sween

Fig. 46 Distinctive stratification on the coast at Mallaig

Fig. 47 A limestone pavement at Aysgarth in the Yorkshire Dales

Fig. 48 Flint nodules in the limestone at Ashford in the Peak District

Fig. 49 The gentle hills at Crowden in the Peak District

Fig. 50 Typical Peak District landscape at Holmsfirth

Fig. 51 Mam Tor, the shivering mountain

Fig. 52 Boulders and pebbles of all sizes on the beach at Whitby

Fig. 53 Interesting sandstone formations at Budleigh Salterton

Fig. 54 The red sandstone cliffs at Sidmouth on the south coast

Bibliography

Bibliography

The research for this book was undertaken primarily via the author visiting a large number of sites throughout Britain and thus gaining first hand experience of the local geology and overall current situation. This approach also facilitated the development of a uniquely up to date geoscience image library, a very small portion of which is reproduced within the pages of this book. Additional research was undertaken via the Internet in order to confirm the historical context and various local factors. The list of books for suggested reading and the web sites visited are described below.

Suggested Reading

British Geological Survey (1971) Northern England. British Geological Survey, United Kingdom

British Geological Survey (1985a) South west England. British Geological Survey, United Kingdom

British Geological Survey (1985b) The midland valley of Scotland. British Geological Survey, United Kingdom

British Geological Survey (1990) The northern highlands of Scotland. British Geological Survey, United Kingdom

Carson (1951) The sea around us. Oxford University Press, USA

Chatwin (1961) East Anglia and adjoining areas. British Geological Survey, United Kingdom

Coenraads et al (2007) Geologica. Millennium House, Australia

Earp (1971) The Welsh borderland. British Geological Survey, United Kingdom

Edwards (1954) The pennines and adjacent area. British Geological Survey, United Kingdom

Hains (1969) Central England. British Geological Survey, United Kingdom

Kibblewhite et al (2005) Soil atlas of Europe. European Commission, Luxembourg

Melville (1983) The Hampshire basin and adjoining areas. British Geological Survey, United Kingdom

Miller (1983) Continents in collision. Time-Life Books, USA

Palmer (2000) The Atlas of the Prehistoric World. Marshall Publishing, United Kingdom

Redfern (2000) Origins. Cassell & Co, United Kingdom

Sumbler (1996) London and the Thames valley. British Geological Survey, United Kingdom

Watson et al (2007) Astronomica. Millennium House, Australia

Associated Web Sites

A taste of the Peak District http://www.thepeakdistrict.info/fast/html/geology_peak_district. Last accessed 19/02/2009.

About geodiversity http://www.geologywales.co.uk/central-wales-rigs. Last accessed 28/04/2009.

Around West Bay – West Dorset cliffs http://www.jurassiccoast.com/277/the-coast-uncovered-30/explore-the-coast. Last accessed 30/12/2008.

Arran – Hutton's unconformity http://www.scottishgeology.com/outandabout/classic_sites/locations/arran. Last accessed 23/01/2009.

Axmouth to Lyme Regis – the undercliff http://www.jurassiccoast.com/277/the-coast-uncovered-30/explore-the-coast. Last accessed 30/12/2008.

Beer and Seaton – building from stone http://www.jurassiccoast.com/277/the-coast-uncovered-30/explore-the-coast. Last accessed 30/12/2008.

British Geological Survey. Coastal erosion at Happisburgh, Norfolk http://www.bgs.ac.uk/landslides/Happisburgh. Last accessed 14/07/2009.

British Geological Survey. Landslides and coastal erosion at Aldbrough http://www.bgs.ac.uk/science/landUseAndDevelopment/landslides/aldbrough. Last accessed 27/08/2009.

British Geological Survey. Landslides and coastal erosion at Sidestrand, Norfolk http://www.bgs.ac.uk/landslides/sidestrand. Last accessed 27/08/2009.

Cairngorms http://www.scottishgeology.com/outandabout/classic_sites/locations/cairngorms. Last accessed 23/01/2009.

Caithness http://www.scottishgeology.com/outandabout/classic_sites/locations/caithness. Last accessed 23/01/2009.

Callander http://www.scottishgeology.com/outandabout/classic_sites/locations/callander. Last accessed 23/01/2009.

J. Ashbourn, *Geological Landscapes of Britain*,
DOI 10.1007/978-90-481-8861-1, © Springer Science+Business Media B.V. 2011

Chalk of Kent http://www.geologyshop.co.uk/chalk1. Last accessed 15/07/2009.

Cleveland hills http://www.visitnorthyorkshiremoors.co.uk/content. Last accessed 30/05/2009.

Coastal erosion unavoidable, EA chief warns http://www.politics.co.uk/news/environment-and-rural-affairs. Last accessed 27/08/2009.

Coastal villages to be sacrificed to the sea http://scratbycoastalerosion.org.uk. Last accessed 27/08/2009.

Craven Dales http://www.yorkshiredales.org.uk/index/learning_about/landscape_character. Last accessed 01/12/2008.

Cumbrian Dales http://www.yorkshiredales.org.uk/index/learning_about/landscape_character. Last accessed 01/12/2008.

Daeareg Ynis Mon http://www.kabrna.com/cpgs/anglesey. Last accessed 07/04/2009.

Earliest evidence found of settlers in Scotland http://www.britarch.ac.uk/ba/ba60. Last accessed 22/12/2008.

Eathie http://www.scottishgeology.com/outandabout/classic_sites/locations/eathie. Last accessed 23/01/2009.

Environmental change and mineral formation in Wales http://www.wales-underground.org.uk/geo. Last accessed 07/04/2009.

Erosion of the Holderness coast http://www.herb.hull.ac.uk/Erosion. Last accessed 27/08/2009.

Geological features http://www.exmoor-nationalpark.gov.uk/index/learning_about/looking_after_landscape. Last accessed 26/01/2009.

Geological history http://www.visitnorthyorkshiremoors.co.uk/content. Last accessed 30/05/2009.

Geology – coal mining http://www.wales-underground.org.uk/pit/geology. Last accessed 28/04/2009.

Geology and Coniston http://www.ruskinmuseum.com/geology. Last accessed 02/11/2009.

Geology and Landforms http://www.dartmoor-npa.gov.uk. Last accessed 26/01/2009.

Geology and mining in the Peak District http://www.buxtononline.net/Buxton-Online98. Last accessed 19/02/2009.

Geology in the North Yorkshire Moors national park http://www.north-york-moors.co.uk/geology. Last accessed 30/05/2009.

Geology of Anglesey: A journey through time http://www.geomon.org.uk/geology2. Last accessed 07/04/2009.

Geology of Fforest Fawr http://www.breconbeacons.org/geopark/understanding/rocks-and-geology-of-the-geopark/geology-of-fforest-fawr. Last accessed 23/01/2009.

Geology of Hertfordshire http://www.hertsgeolsoc.ology.org.uk. Last accessed 14/07/2009.

Geology of Kent and the Boulonnais http://www.geologyshop.co.uk/geolkb. Last accessed 15/07/2009.

Geology of north east Wales http://www.newrigs.org/upal_eng. Last accessed 07/04/2009.

Geology of the local area http://www.great-bradley.suffolk.gov.uk/History/landscape. Last accessed 14/07/2009.

Happisburgh's problems are national problems http://www.happisburgh.org.uk. Last accessed 29/10/2009.

Introduction to Exmoor's archaeologyhttp://www.exmoor-nationalpark.gov.uk/index/learning_about/looking_after_landscape. Last accessed 26/01/2009.

Islands and the Neolithic farming tradition http://www.britarch.ac.uk/ba/ba71. Last accessed 22/12/2008.

Isle of Skye – The Cuillin Hills http://www.scottishgeology.com/outandabout/classic_sites/locations/skye_cuilin_hills. Last accessed 23/01/2009.

Lake District fells http://www.lakedistrict-guide.co.uk/lakedistrict_fells. Last accessed 02/11/2009.

Lake District Geology http://nvmdigital.net/topics/Geology/geology. Last accessed 02/11/2009.

Lynmouth area http://www.devon.gov.uk/index/environment/natural_environment/geology. Last accessed 26/01/2009.

Mesolithic and Neolithic settlers http://www.merseyreporter.com/history/historic/mesolithic. Last accessed 22/12/2008.

Middle Lias of Staithes, Yorkshire http://www.soton.ac.uk/~imw/staithes. Last accessed 30/05/2009.

Minerals Local Plan – Geology of Cornwall http://www.cornwall.gov.uk. Last accessed 05/01/2009.

On the rocks http://www.bbc.co.uk/wales/northwest/sites/nature/pages. Last accessed 07/04/2009.

Peak District geology http://www.peakscan.freeuk.com/peak_district_geology. Last accessed 19/02/2009.

Pennine Way http://www.cressbrook.co.uk/outdoors/pennineway. Last accessed 02/04/2009.

Regional geology – Hebrides http://www.scottishgeology.com/geology/regional_geology/hebrides. Last accessed 23/01/2009.

Regional geology – Orkney, Shetland http://www.scottishgeology.com/geology/regional_geology/northern_isles. Last accessed 23/01/2009.

Scotland's first settlers http://www.articlearchives.com/environment-natural-resources/ecology. Last accessed 22/12/2008.

Searching for the first settlers of Orkney http://www.oat.org.uk/dhl/papers/cwj. Last accessed 22/12/2008.

Southern Howgill fells http://www.yorkshiredales.org.uk/index/learning_about/landscape_character. Last accessed 01/12/2008.

Sutton Bank http://www.visitnorthyorkshiremoors.co.uk/content. Last accessed 30/05/2009.

Swaledale and Arkengarthdale http://www.yorkshire-dales.org.uk/index/learning_about/landscape_character. Last accessed 01/12/2008.

The Derbyshire Peak District http://www.brixworth.demon.co.uk/peak. Last accessed 19/02/2009.

The disappearing East Anglian coastline http://www.marinet.org.uk/mad/disappearingcoastline. Last accessed 14/07/2009.

The Formation of Tors on the Dartmoor Granite – an outline http://www.dartmoor-npa.gov.uk. Last accessed 26/01/2009.

The geology of the Abercrave area http://www.wales-hotel.co.uk/geology. Last accessed 28/04/2009.

The geology of White Colne http://www.whitecolne-history.co.uk. Last accessed 15/07/2009.

The Heritage Journal. Coastal erosion – Skara Brae http://heritageaction.wordpress.com/2009/08/04/coastal-erosion-skara-brae. Last accessed 27/08/2009.

The Lizard http://en.wikipedia.org/wiki/The_Lizard. Last accessed 05/01/2009.

The Management of Coastal Erosion – Report 1 http://www.canterbury.gov.uk. Last accessed 27/08/2009.

Wensleydale http://www.yorkshiredales.org.uk/index/learning_about/landscape_character. Last accessed 01/12/2008.

Index

J. Ashbourn, *Geological Landscapes of Britain*,
DOI 10.1007/978-90-481-8861-1, © Springer Science+Business Media B.V. 2011